"Amos Oz is one of Israel's most prolific, celebrated writers, capturing the past and exploring the present in more than thirty novels, dozens of essays, and hundreds of articles. But his latest book, *Dear Zealots: Letters from a Divided Land*, may contain his most urgent message yet."

— RUTH EGLASH, *WASHINGTON POST*

"Three passionate lectures about the state of politics in Israel. In this rumination about the country he loves and whose policies make him ashamed, novelist and peace activist Oz sounds humorous, mournful, enraged, and uplifting . . . Oz maintains there's rarely been a better moment to make peace than now . . . Slender but forceful."

— *KIRKUS REVIEWS*

"Concise, evocative . . . *Dear Zealots* is not just a book of thoughts and ideas — it is a depiction of one man's struggle, who for decades has insisted on keeping a sharp, strident, and lucid perspective in the face of chaos and at times of madness."

— DAVID GROSSMAN

"Readers unfamiliar with Israeli author and public intellectual Oz will find this collection of three essays, adapted from a series of lectures, a good introduction to his nuanced perspective . . . Clear-eyed . . . Providing a worthy companion volume to Yossi Klein Halevi's *Letters to My Palestinian Neighbor*, Oz's book leaves readers with a strong message about the need for a greater society-wide openness to doubt and ambiguity."

— *PUBLISHERS WEEKLY*, STARRED REVIEW

"Celebrated Israeli novelist Oz writes nonfiction, too, including the three essays collected here, relevant to our polarized, populist world: they treat the nature of fanaticism, the Jewish roots of humanism (and the need for a secular appreciation of Israel), and Israel's geopolitical standing. Oz says he wrote them for his grandchildren, but they're good for us all."

— BARBARA HOFFERT, *LIBRARY JOURNAL*

DEAR ZEALOTS

Books by Amos Oz

FICTION

Where the Jackals Howl

Elsewhere, Perhaps

My Michael

Unto Death

Touch the Water, Touch the Wind

The Hill of Evil Counsel

A Perfect Peace

Black Box

To Know a Woman

Fima

Don't Call It Night

Panther in the Basement

The Same Sea

Rhyming Life and Death

Scenes from Village Life

Between Friends

Judas

AMOS OZ

DEAR ZEALOTS

Letters from a
Divided Land

TRANSLATED FROM THE HEBREW
BY JESSICA COHEN

Mariner Books
Houghton Mifflin Harcourt
BOSTON   NEW YORK

First Mariner Books edition 2019
Copyright © 2017 by Amos Oz
Translation copyright © 2018 by Jessica Cohen

First published in Hebrew as *Shalom la-Kana'im*

hmhbooks.com

Library of Congress Cataloging-in-Publication Data
Names: Oz, Amos, author. | Cohen, Jessica (Translator)
Title: Dear zealots : letters from a divided land /
Amos Oz ; translated from the Hebrew by Jessica Cohen.
Other titles: Shalom la-Kana'im Hebrew
Description: Boston : Houghton Mifflin Harcourt, [2018] |
Includes bibliographical references.
Identifiers: LCCN 2018017516 (print) | LCCN 2018018571 (ebook) |
ISBN 9781328987563 (ebook) | ISBN 9781328987006 (hardcover) |
ISBN 9780358175445 (paperback) |
Subjects: LCSH: Fanaticism. | Judaism and humanism. |
Democracy — Religious aspects — Judaism. | Toleration.
Classification: LCC BF575.F16 (ebook) | LCC BF575.F16 O913 2018 (print) |
DDC 892.4/6 — dc23
LC record available at https://lccn.loc.gov/2018017516

Book design by Carly Miller

Printed in the United States of America
DOC 10 9 8 7 6 5 4 3 2 1

"The Place Where We Are Right" by Yehuda Amichai, from *The Selected
Poetry of Yehuda Amichai*, edited and translated from the Hebrew by Chana
Bloch and Stephen Mitchell. © University of California Press, 1996.

The essay "Dear Zealots" is based on a series of lectures delivered at the
University of Tübingen in Germany, in 2002, and subsequently published in
How to Cure a Fanatic, a small book that was translated into more than twenty
languages. The essay appears here in an expanded and updated version.

The essay "Many Lights, Not One Light" derives from the book *Jews and Words*,
which I wrote with my daughter Fania Oz-Salzberger in 2012 (Yale University
Press). It is also based on a lecture titled "A Full Cart and an Empty Cart," which I
delivered many years ago at Bar-Ilan University and which appeared in a concise
version in my book *All Our Hopes* (Keter, 1998). An additional source was a lecture
I gave during a Shavuot event in 2016 at the Tel Aviv home of the Shenhav family.

The essay "Dreams Israel Should Let Go of Soon" is based on a lecture I gave
at a seminar in memory of Lieutenant General Amnon Lipkin-Shahak at the
Interdisciplinary Center Herzliya, and on a different version of the lecture
delivered at the Institute for National Security Studies (both in 2015).

To my grandchildren,
Dean, Nadav, Alon and Yael, with love and respect.
This book was written, first and foremost, for you.

Contents

Preface

THE THREE ESSAYS THAT FOLLOW WERE written not by a scholar or an expert, but by a person living through and grappling with the situation. The essays' common thread is my desire to take a personal look at a number of extremely controversial issues, some of which strike me as matters of life or death.

This book does not purport to describe every aspect of every disagreement or to elucidate all features of the landscape, and certainly not to have the last word. Rather, it seeks the listening ear of those whose opinions differ from my own.

— AMOS OZ

THE PLACE WHERE WE ARE RIGHT

From the place where we are right
flowers will never grow
 in the spring.

The place where we are right
is hard and trampled
 like a yard.

But doubts and loves
dig up the world
like a mole, a plow.
And a whisper will be heard in the place
where the ruined
 house once stood.

— YEHUDA AMICHAI, TRANSLATED BY
STEPHEN MITCHELL

DEAR
ZEALOTS

I

Dear Zealots

HOW DOES ONE CURE A FANATIC? SETTING off in pursuit of a gang of armed zealots in the mountains of Afghanistan, the deserts of Iraq, or the cities of Syria is one thing. Fighting zealotry itself is quite another. I have nothing new to suggest regarding desert and mountain wars, or their online counterparts. But here are a few thoughts about the nature of fanaticism and the ways we might curtail it.

The attack on the Twin Towers in New York, on September 11, 2001, much like dozens of attacks on urban centers and bustling sites around the world, did not stem from the poor being angry at the rich. Wealth disparity is an age-old injustice, but the new wave of violence is not solely, or primarily, a response to that disparity. If it were, the onslaught of terrorism would have originated in African countries — the poorest — and landed in Saudi Arabia and the Gulf states — the wealthiest. This war is being fought between fanatics convinced that their ends sanctify all means, and everyone else — all those who hold that life is an end and not a means. It is a struggle between people who believe that justice, whatever that term may mean to them, is more important than life, and those who maintain that life takes precedence over other values.

EVER SINCE THE POLITICAL SCIENTIST Samuel Huntington defined the current worldwide battlefield as a "war of civilizations" being fought primarily between Islam and Western civilization, the prevalent worldview has been a racist picture that portrays a clash between "savage terrorist" Easterners and "cultured" Westerners. This was not Huntington's formula, but such is the predominant sentiment aroused by his writings.

The Israeli government, for example, finds it convenient to lean on this trite Wild West formula, because it allows it to dump the Palestinians' fight to cast off Israeli occupation into the same junkyard from which fanatic Muslim murderers regularly emerge to commit horrors around the world.

Many people forget that radical Islam does not have a monopoly on violent fanaticism. The destruction of the Twin Towers and the bloodshed that continues in various parts of the world are not necessarily tied to the questions: Is the West good or bad? Is globalization a blessing or a monster? Is capitalism loathsome or self-evident? Are secularism and hedonism enslavement or freedom? Is Western colonialism over and done with or has it simply taken on new forms? These questions can have different, and even contradictory, answers, without any of them being fanatic. The fanatic does not argue. If something is wrong in his view, if it is clear to him that something is wrong in God's view, it is his duty to destroy the abomination, even if that means killing anyone who just happens to be around.

• • •

FANATICISM DATES BACK MUCH EARLIER than Islam. Earlier than Christianity and Judaism. Earlier than all the ideologies in the world. It is an elemental fixture of human nature, a "bad gene." People who bomb abortion clinics, murder immigrants in Europe, murder Jewish women and children in Israel, burn down a house in the Israeli-occupied territories with an entire Palestinian family inside, desecrate synagogues and churches and mosques and cemeteries — they are all distinct from al-Qaeda and ISIS in the scope and severity of their acts, but not in their nature. Today we speak of "hate crimes," but perhaps a more accurate term would be "zealotry crimes," and such crimes are carried out almost daily, including against Muslims.

Genocide and jihad and the Crusades, the Inquisition and the gulags, extermination camps and gas chambers, torture dungeons and indiscriminate terrorist attacks: none of these are new, and almost all of them preceded the rise of radical Islam by centuries.

As the questions grow harder and more complicated, people yearn for simpler answers, one-sentence answers, answers that point unhesitatingly to a culprit who can be blamed for all our suffering, answers that promise that if we only eradicate the villains, all our troubles will vanish.

"It's all because of globalization!" "It's all because of the Muslims!" "It's all because of permissiveness!" or "because of the West!" or "because of Zionism!" or "because of immigrants!" or "because of secularism!" or "because of the left wing!" All one needs to do is cross out the incorrect entries, circle the right Satan, then kill that Satan (along with his neighbors and anyone who happens to be in the area), thereby opening the gates of heaven once and for all.

More and more commonly, the strongest public sentiment is one of profound loathing — subversive loathing of "the hegemonic discourse," Western loathing of the East, Eastern loathing of the West, secular loathing of believers, religious loathing of the secular. Sweeping, unmitigated loathing surges like vomit from the depths of this or that misery. Such extreme loathing is a component of fanaticism in all its guises.

For example, concepts that only half a century ago seemed innovative and exciting — multiculturalism and identity politics — quickly morphed, in many places, into the politics of identity hatred. What began with an expansion of cultural and emotional horizons is increasingly deteriorating into narrower horizons, isolationism, and hatred of the other. In short, a new wave of loathing and extremism assails us from all sides.

PERHAPS MY CHILDHOOD IN JERUSALEM gave me some expertise in comparative fanaticism. In the 1940s there was no shortage of open hearts and capacious souls in Jerusalem. But there were also a lot of self-appointed prophets, saviors, and messiahs. To this day, almost every second or third Jerusalemite has his or her own private formula for instant salvation. Of course, lots of people claim to be in Jerusalem in order "to build and be built in it," as the old Zionist song would have it, but many of them — Jews, Muslims, Christians, revolutionaries, radicals, world-healers — came to Jerusalem not to build and be built, but to crucify or be crucified.

A well-known mental illness has garnered the medical diagnosis "Jerusalem syndrome": no sooner do people breathe in the crisp mountain air ("clear as wine," in the words of one famous Hebrew song) than they set off to burn a mosque or blow up a church or destroy a synagogue, to kill heretics or believers, to "eradicate evil from the world." Most sufferers of Jerusalem syndrome, however, make do with stripping down, climbing atop a rock, and prophesizing.

Although they have few believers, these prophets are numerous, spanning from one end of the spectrum to the other. Their common denominator is an urge to fulfill a simple salvation formula, and sometimes to point to the villains from which the world must be purified in order to hasten redemption. Redemption itself, according to most of these prophets, is easily crammed into a short slogan.

· · ·

AS A CHILD IN JERUSALEM, I MYSELF WAS a little Zionist-nationalist fanatic—enthusiastic, self-righteous, and brainwashed. I was blind to any argument that deviated at all from the Jewish-Zionist story we were told by nearly all the adults around us. I was deaf to any reasoning that challenged that story. Like all the children in the Kerem Avraham neighborhood, I also threw stones at the British vehicles that patrolled our little street. Along with the stones, we hurled almost our entire English vocabulary at the soldiers: "British, go home!" All this happened in 1946 or 1947, as the British Mandate over Palestine was nearing its end, in the days of the original intifada—the one instigated by us Jews against the British occupation. This, too, I suppose, is another example of the irony of history.

IN MY NOVEL *PANTHER IN THE BASEMENT*, I retold the experiences that revealed to me, as a child, that sometimes there are two sides to a story, that conflicts are colored not only in black and white. In the last year of the British Mandate, when I was about eight, I befriended a British policeman who spoke ancient Hebrew and had memorized most of the Bible. He was a fat, asthmatic, emotional man, and perhaps a slightly muddled one, who fervently believed that the Jewish people's return to its ancestral land heralded redemption for the world at large. When the other children discovered my friendship with this man, they called me a traitor. Much later, I learned to take comfort in the thought that, for fanatics, a traitor is anyone who dares to change. Fanatics of all kinds, in all places at all times, loathe and fear change, suspecting that it is nothing less than a betrayal resulting from dark, base motives.

The boy narrator in *Panther in the Basement* begins the story as a Zionist fanatic, afire with his sense of righteousness, but within weeks he learns, to his astonishment, that there are things in the world that can be seen one way but can also be seen in a completely different way. This discovery signals the character's loss of innocence, but he gains a larger world, along with a certain expertise in comparative fanaticism. He comes to learn that blind hatred often turns the haters on either side of the fence into almost identical personas.

The term "comparative fanaticism" is no joke. Perhaps the time has come for every university, school, and educational institution to introduce a course or two in comparative fanaticism, because fanaticism is closing in on us, here in Israel and in many other places in the world, east and west, north and south. And we are not talking only of extreme Islam: in many places, at this very moment, there are dangerous waves of Christian religious fanaticism (in the United States, Russia, and a few Eastern European countries), murky tides of Jewish religious fanaticism, dark surges of isolationist and xenophobic nationalism in Western and Eastern Europe, and an increasing flood of racism in more and more societies.

The fanaticism in almost all of Jewish-Israeli society, of various shades and types, arrived in Israel with the Jews of Europe. From Eastern Europe we received the revolutionary fanaticism of the founding pioneer generation, bent on remolding the Jewish people and erasing the abundance of heritages of those who came from other diasporas in order to rebuild a "new Hebrew." Europe was also the source of our nationalist fanaticism, with its worship of militarism and all sorts of delusions of imperialist grandeur. Also from Europe: ultra-Orthodox fanaticism, which secludes itself inside a walled ghetto and defends itself against anything different. Jewish immigrants from Middle Eastern and North African countries, conversely, brought a generations-old heritage of moderation, relative religious tolerance, and the custom of living in good neighborly relations even with those who are different. Indeed, the various forms of European fanaticism are now erasing the moderation of Middle Eastern Jews.

One of the reasons for the increasing waves of fanaticism, then, might be the growing desire for simple, decisive solutions, for "one-shot salvation." Another reason is that we are increasingly distancing ourselves from the horrors that occurred in the first half of the twentieth century: Stalin and Hitler, unintentionally, seem to have invested the two or three generations that followed with a

profound fear of any extremism and with a measure of restraint toward fanatical urges. For a few
decades, thanks to the greatest murderers of the
twentieth century, racists were a little bit ashamed
of their racism, haters moderated their hatred
slightly, and fanatic world-healers were careful
not to get too revolutionary. Perhaps not everywhere, but at least here and there.

In recent years, this "gift" from Stalin, from Hitler, from the Japanese militarists, is approaching
its expiration date. The partial immunity we absorbed is fading. Hatred, zealotry, loathing of the
other and those who are different, revolutionary
murderousness, the zeal to crush all the villains in
a bloodbath — all are rearing their heads again.

FANATICISM IS NOT RESERVED FOR AL-
Qaeda and ISIS, Jabhat al-Nusra, Hamas and Hezbollah, neo-Nazis and anti-Semites, white su-
premacists and Islamophobes and the Ku Klux
Klan, Israel's "hilltop thugs" in the settlements,
and others who would shed blood in the name
of their faith. These fanatics are familiar to us

all. We see them every day on our television screens, shouting, waving angry fists at the camera, hoarsely yelling slogans into the microphone. They are the visible fanatics. A few years ago, my daughter Galia Oz directed a documentary film that probed the roots of fanaticism and its manifestations in the Jewish underground.

But there are far less prominent and less visible forms of fanaticism around us, and perhaps inside us, too. Even in the daily lives of normative societies and people we know well, there are sometimes revelations, albeit not necessarily violent ones, of fanaticism. One might encounter, for example, fanatic opponents of smoking who act as if anyone who dares light a cigarette near them should be burned alive. Or fanatic vegetarians and vegans who sometimes sound ready to devour people who eat meat. A few of my friends in the peace movement denounce me furiously, simply because I hold a different view of the best way to achieve peace between Israel and Palestine.

Certainly, not everyone who raises a voice for or against something is suspected of fanaticism, and not everyone who angrily protests an injustice becomes a fanatic by virtue of that protest and anger. Not every person with strong opinions is guilty of fanatic tendencies. Not even when such views or emotions are expressed very loudly. It is not the volume of your voice that defines you as a fanatic, but rather, primarily, your tolerance — or lack thereof — for your opponents' voices.

Indeed, a hidden — or not so hidden — kernel of fanaticism often lies beneath various disclosures of uncompromising dogmatism, of imperviousness and even hostility toward positions you deem unacceptable. Righteousness entrenched and buttressed within itself, righteousness with no windows or doors, is probably the hallmark of this disease, as are positions that arise from the turbid wellsprings of loathing and contempt, which erase all other emotions. (There is nothing wrong with loathing in and of itself: in Shakespeare and Dostoyevsky and Brecht, Chaim Nachman Bialik and Y. H. Brenner and Hanoch Levin, we find a stinging component of loathing. A blazing component — but not an exclusive one. In the works of these great writers, loathing is accompanied by other feelings, too — by understanding, compassion, longing, humor, and a measure of sympathy.)

THERE ARE VARYING DEGREES OF EVIL IN the world. The distinction between levels of evil is perhaps the primary moral responsibility incumbent upon each of us. Every child knows that cruelty is bad and contemptible, while its opposite, compassion, is commendable. That is an easy and simple moral distinction. The more essential and far more difficult distinction is the one between different shades of gray, between degrees of evil. Aggressive environmental activists, for example, or the furious opponents of globalization, may sometimes emerge as violent fanatics. But the evil they cause is immeasurably smaller than that caused by a fanatic who commits a large-scale terrorist attack. Nor are the crimes of the terrorist fanatic comparable to those of fanatics who commit ethnic cleansing or genocide.

Those who are unwilling or unable to rank evil may thereby become the servants of evil. Those who make no distinction between such disparate phenomena as apartheid, colonialism, ISIS, Zionism, political incorrectness, the gas chambers, sexism, the 1 percent's wealth, and air pollution serve evil with their very refusal to grade it.

Fanatics tend to live in a black-and-white world, with a simplistic view of good against evil. The fanatic is in fact a person who can only count to one. Yet at the same time, and without any contradiction, the fanatic almost always basks in some sort of bittersweet sentimentalism, composed of a mixture of fury and self-pity. He or she prefers to feel instead of think. Death — their own or someone else's — enthralls fanatics and excites their imagination. Not infrequently, they find this world despicable and loathsome, and aspire to escape it as soon as possible (and take as many people with them as they can). The fanatic longs to replace this evil world with a better one, an "afterworld," with or without the seventy-two virgins who await as reward for his sacrifice.

· · ·

CONFORMISM, BLINDLY TOEING THE LINE, obedience without contemplating or questioning, the common desire to belong to a tightly unified human bloc — these, too, are cornerstones of the fanatic soul. In *Monty Python's Life of Brian,* the redeemer Brian preaches to his crowd: "You are all individuals!" The crowd roars back: "Yes! We are all individuals!" "You are all different!" he continues to rile them up. "Yes! We are all different!" But one little man softly bleats: "I'm not."

The urge to follow the crowd and the passion to belong to the majority are fertile ground for fanatics, as are the various cults of personality, idolization of religious and political leaders, and the adulation of entertainment and sports celebrities.

Of course there is a great distance between blindly worshiping bloodthirsty tyrants, being swept up by murderous ideologies or aggressive, hateful chauvinism, and the inane adoration of celebrities. Still, there is perhaps a common thread: the worshiper yields his own selfhood. He longs to merge — to the point of self-deprecation — with the throng of other admirers and unite with the experiences and accomplishments of the object of worship. In both cases, the elated admirer is subjugated by a sophisticated system of propaganda and brainwashing, a system that intentionally addresses the childish element in people's souls, the element that so longs to merge, to crawl back into a warm womb, to once again be a tiny cell inside a huge body, a strong and protective body — the nation, the church, the movement, the party, the team fans, the groupies — to belong, to squeeze in with a crowd under the broad wings of a great father, an admired hero, a dreamy beauty, a sparkling celebrity, in whose hands the worshipers deposit their hopes and dreams, and even their right to think and judge and take positions.

THE INCREASING INFANTILIZATION OF masses of people everywhere in the world is no coincidence: there are those who stand to gain from it and those who ride its coattails, whether from a thirst for power or a thirst for wealth. Advertisers and those who fund them desperately want us to go back to being spoiled little children, because spoiled little children are the easiest consumers to seduce.

The boundary between politics and entertainment is being erased before our eyes. The world is turning into a kind of global kindergarten. The qualities a candidate needs to be elected are antithetical to the qualities required to lead. Politics and the media have become branches of the entertainment industry: like in ancient Rome, the media throws two or three famous victims to the lions every day, whether innocent or guilty, to entertain the masses, distract them, and extract their money.

Growing segments of the public vote for the candidate who can excite and amuse them, the one who snubs the acceptable rules of the game. More and more people vote for the heck of it, or for kicks, or for laughs. Ostensibly, this is all radically different from the gloomy solemnity of fanatics marching in formation, eyes shut, to the sound of a charismatic leader's drumbeat. But the dissimilarity is merely superficial: the fanatic may al-

ways be deathly serious, but the juvenile follow-
ers who are out for laughs and kicks are in just as
much of a hurry to relinquish their right to a rich
and diverse life. Both the marchers and the jok-
ers are throwing away their freedom to think, to
consider, to make choices, and to sometimes vary
their choices.

ONE OF THE DISTINCT HALLMARKS OF THE
fanatic is his fervent desire to change you so that
you will be like him. To convince you that you
must immediately convert, abandon your world,
and move into his. The fanatic does not want
there to be any differences between people. He
wants us all to be as one. He desires a world with
no curtains drawn, no blinds shuttered, no doors
locked, no shadow of a private life, for we must all
be one body and one soul. We must all march together in threes on the path ascending to redemption, whether this redemption or the opposite one.

The fanatic strives to upgrade and improve
you, to open your eyes so that you, too, can see the
light. Indeed, in that sense the fanatic is a won-

drously altruistic and extremely unselfish crea-
ture: he is interested in you far more than he is in
himself. Day and night he yearns to save your soul,
to unshackle you, to take you out of darkness into
the light, to redeem you once and for all from error and sin. Here he comes to hug you, sick with
worry about your condition, bubbling with good-
will to change your prayer habits (or lack thereof),
your voting or smoking habits, your eating habits, your preferences, your entire lifestyle, which is
so harmful to you. All the fanatic wants is to take
you in his arms and hug you, to raise you from the
lowly spot you are stuck in and place you in the
sublime place he has discovered, where he has
since been basking and to which you must ascend
immediately. For your very own good.

The fanatic is always in a hurry to fall on your
neck to save you, because he loves you. He loves
you unconditionally. But, conversely, he might
grab you and strangle you if he discovers that you
are beyond redemption. Lost. And if that is the
case, he is obliged to hate you and rid the world of
you.

The reason the fanatic is far more interested in you than in himself is that he usually does not have a self, or hardly any. The fanatic is a wholly public persona. He has no private life. And if he does, he flees it. Winston Churchill defined the fanatic as "one who can't change his mind and won't change the subject." It still holds true.

RELIGIOUS FANATICS AND IDEOLOGICAL fanatics of various strains commit horrific crimes of violence not only because they loathe the heretics, or the West, or Muslims, or lefties, or Zionists, or LGBTQ people. They shed blood because they want to save the world. To save the heretics from themselves, to extricate them from the depths of their heresy. To put us all on the straight and narrow path. To permanently remove us, by blood and fire, from our noxious values.

To a fanatic Muslim, for example, materialism is a mental illness. Permissiveness is a brutish phenomenon. Enslavement to gadgets is a despicable mental perversion, as are women's liberation and tolerance of different sexual identities. Addiction to drugs and pornography not only infuriates God but destroys the lives of the addicts themselves. Decadent democracy, repulsive pluralism, scandalous freedom of expression — to a fanatic Muslim and to all other religious fanatics these are but temptations held out by Satan to make us desecrate God's laws. And they, the fanatic warriors, have been sent by God to purify the world, to rid us of the filth that clings to us. They have an urgent duty to open the eyes of moderate believers who dare to compromise their complex reality, thereby betraying God and the faith. To Muslim fanatics, as to other religious fanatics, the only medicine for all human diseases is to accept religious laws in their strictest version.

Apart from extreme religious believers, there are other fanatics who strive to change our confused humanity in order to save it from itself, and they include groups of radical world-healers and violent revolutionaries.

The fanatic is convinced that it is not enough to deliver bloody blows to "heretics" and "deviants." The bloody blows are merely a regretful but necessary step on the way to illuminating for us the path to the only faith and the sole salvation. In short: ISIS loves you profoundly. Al-Qaeda is devoted to saving you from your moral deterioration. The Ku Klux Klan, or Lehavah (a far-right vigilante group in Israel that opposes "assimilation" of Jews), or the "hilltop" settler youth gangs endanger their lives and sacrifice themselves not on their own behalf but on yours. They are here to redeem you, even if at present you are still blinded and cannot see what is best for you. They were sent by God or by the ideology to offer you a lifeline. If you refuse to grasp it, they will be forced to hurt you. The day will come when your eyes will be opened, you will see the light, and you will be grateful to them. When that day comes, you will fall to your knees and thank them for saving you from yourself, because the fanatic will always see himself as the parent and you, to use the Talmudic term, as a "captured infant."

THE ILLNESS FREQUENTLY BEGINS WITH innocuous symptoms: not beheadings, not car bombs, not burning families alive in their homes, but rather in the bosom of the family. Fanaticism begins at home. Its milder manifestations, which we all know, are expressed in the ubiquitous urge to change, just slightly, your beloveds, your children, your siblings, your partner, your neighbors — to change them for their own good. Sometimes even to sacrifice yourself for them, especially when they are utterly blind and simply do not grasp the evil they are inflicting upon themselves every day. "Be like me." "Learn from me." These daily phrases can carry a hidden load of fanaticism.

Sometimes, though, fanaticism is born from a desire to live your own life through someone else's: that of a prophet or a miracle maker, a political leader or a glamour girl, a famous artist or an athlete. In extreme cases, this urge may go as far as the willingness to completely efface yourself (or others) in order to earn the recognition and blessing of your heroes.

At times the fanatical urge travels a different channel: "We would happily sacrifice ourselves for the future of our children," or "for future generations," or "to hasten the coming of the Messiah," or "to mend the world," or "to make God happy," or "for redemption," or whatever else. In many cases, the child whose parents "sacrificed themselves for him" is destined to bear a heavy burden of guilt for the rest of his or her life. Between the two mothers, as the famous joke goes, the one who roars at her child, "If you don't finish your oatmeal I'll kill you!" and the one who wails, "Finish your oatmeal or I'll kill myself!"—many would choose the former, which is indeed the lesser evil. Self-sacrifice does not always represent an erasing of the "I" in favor of something more precious than "I." Self-sacrifice can sometimes be a well-honed weapon that the fanatic wields for destructive emotional purposes.

Moreover, those who are eager to sacrifice themselves will not find it difficult to sacrifice others.

A PROMINENT ISRAELI WRITER, SAMI MIchael, once told of a long car journey with a driver. At some point, the driver explained to Michael how important, indeed how urgent, it is for us Jews "to kill all the Arabs." Sami Michael listened politely, and instead of reacting with horror, denunciation, or disgust, he asked the driver an innocent question: "And who, in your opinion, should kill all the Arabs?"

"Us! The Jews! We have to! It's either us or them! Can't you see what they're doing to us?"

"But who, exactly, should actually kill all the Arabs? The army? The police? Firemen, perhaps? Or doctors in white coats, with syringes?"

The driver scratched his head, pondered the question, and finally said, "We'll have to divvy it up among us. Every Jewish man will have to kill a few Arabs."

Michael did not let up: "All right. Let's say you, as a Haifa man, are in charge of one apartment building in Haifa. You go from door to door, ring the bells, and ask the residents politely, 'Excuse me, would you happen to be Arabs?' If the answer is yes, you shoot and kill them. When you're done killing all the Arabs in the building, you go downstairs and head home, but before you get very far you hear a baby crying on the top floor. What do you do? Turn around? Go back? Go upstairs and shoot the baby? Yes or no?"

A long silence. The driver considers. Finally he says, "Sir, you are a very cruel man!"

This story exposes the confusion sometimes found in the fanatic's mind: a mixture of intransigence with sentimentality and a lack of imagination. Sami Michael, by invoking the baby, forced the fanatic at the wheel to use his imagination, and so strummed on the emotional string in his soul. The baby-loving driver became confused, hurt, and filled with anger at the passenger who had forced him to define, by means of a horrific image, the simpleminded slogan "Death to Arabs!" But it is in the driver's fury, perhaps, that we find a small glimmer of hope, albeit partial and tentative: when the fanatic is forced to visualize the slogan, to imagine the details of the horror and find himself in the role of a baby murderer, perhaps sometimes — only sometimes — he will feel a certain embarrassment. A slight hesitation. A tiny crack will surface in the wall of imperviousness.

Certainly, this is no magic pill. Nevertheless,

perhaps activating one's imagination, being forced to look at the suffering of one's victims at close range, may have the power, here and there, to act as an antidote to the simplified cruelty of slogans such as "Death to Arabs!" and "Death to Jews!" and even "Death to fanatics!" Killing all the Arabs is a lot simpler than killing one Arab baby, one thirsty baby who grabs your arm, sobs, and begs for water because her mouth is dry.

CURIOSITY AND IMAGINATIVE POWER: these two things may give us partial immunity to fanaticism. The story told by Sami Michael, who apparently managed momentarily to embarrass or confuse the driver who advocated killing all Arabs, indicates that the fanatic is uncomfortable imagining the details of the act he eagerly volunteers to perform. He is comfortable with the slogan, as long as the slogan doesn't translate into shouts, pleas, dying gurgles, puddles of blood, brains spilled out on the sidewalk. It is true that there are sadists in the world who would actually be excited by close-up pictures of abuse and dismem-

berment, but most fanatics are not driven by sa-
dism but by lofty ideals, a longing for redemption
and a desire to mend the world, which necessitate
"getting rid of the bad ones."

To imagine the inner world, both intellectual
and emotional, of the other. To use our imagination even in times of strife. To use it also, primarily, in moments when we feel a surge of fury, insult, loathing, righteousness, and the certainty
that we have been wronged and that justice is entirely on our side. Perhaps also to ask, once in a
while: What if I were her? Or him? Or them? To
step, for a moment, into the other's shoes and under his skin, not in order to cross the river or be
"reborn," but simply to understand, to sense, what
is there. What is beyond the river? What do they
have in their head? How do they feel over there?
And what do we look like from there? Perhaps
also to try to find out how deep the dividing river
is. How wide? How and where might we build a
bridge? This curiosity will not necessarily lead us
to a conclusion of sweeping moral relativity, nor
to self-abdication in favor of the other's selfhood.

It will lead us, sometimes, to an exhilarating discovery, which is that there are many rivers, each of whose banks can show us a different landscape that may be fascinating and surprising. Fascinating even if it is not right for us; surprising even if it does not appeal to us. Perhaps, indeed, in curiosity lies the prospect of openness and tolerance.

CURIOSITY AND IMAGINATION ARE BOUND together. The age-old human urge to peek behind other people's drawn shutters, the eagerness to compare one's own intimate secrets with the secret intimacy of others, is an urge that may serve as an antidote to the fanatic's lust to murder the difference between himself and others. Or to kill anyone who refuses to change and declines to be exactly like him.

Literature and gossip are closely related. People who are curious and imaginative long to know "what it's like for other people." This longing can be satisfied in its basest, most banal form through gossip, just as it can attain a more refined and complex gratification in art. Both gossip and literature, each in its own way, are capable of offering a partial antidote to fanaticism, because they both relish the fascinating differences between people.

IN ADDITION TO CURIOSITY AND IMAGINA-tion, another effective antidote to fanaticism might be humor, and especially the ability to make fun of ourselves. I, for one, have never met a fanatic with a sense of humor. Nor have I ever known anyone capable of making a joke at his own expense become a fanatic. Humor engenders a curvature that allows one to see, at least momentarily, old things in a new light. Or to see yourself, at least for a moment, as others see you. This curvature invites us to let the hot air out of any excessive importance, including self-importance. Moreover, humor usually entails a measure of relativity, of abasing the sublime.

If only we could find a way to put a sense of humor, and particularly self-humor, into a capsule and inoculate entire populations against the plague of fanaticism!

Yet how easy it is to fall into such a trap: the

idea of stuffing a sense of humor into a capsule and making people swallow it for their own good, to cure them — the very idea borders on fanaticism. For our humor capsules are based on the assumption that there is someone out there who knows what is good for everyone.

Fanaticism, therefore, is contagious: a person may catch it even as he fights to cure other people of it. There is no shortage of anti-fanaticism fanatics in the world. All sorts of crusades to stop jihad, and jihads to subjugate the new crusaders. This includes the zeal so prevalent in Israel and in the West these days to deliver a knockout blow that will finish off all the bloodthirsty fanatics and anyone like them once and for all. To eradicate every last bastion of zealotry.

IT IS POSSIBLE THAT THE ONLY CONSTITU-ency with the power to halt the rise of extreme Islam is in fact moderate Islam. This is also probably applicable to extremists of other religions and faiths. Violent fanatics must not be allowed to make us forget the simple fact that the overwhelming majority of religious believers in the world, Muslims and others, conduct their daily lives with moderate religiosity that is averse to violence and murder.

Like all types of zealotry, violent Islam is not limited to a gang of sadistic, bloodthirsty fanatics. At its foundation stands an idea. A bitter, desperate idea, a distorted idea. However, it is worth remembering that one can almost never vanquish an idea, twisted as it may be, simply by using a big stick. There must be a response; there must be an opposing idea, a more attractive belief, a more persuasive promise. I am unopposed to using a big stick against murderers. I do not believe in turning the other cheek, nor do I share the prevalent opinion whereby violence is the absolute evil. To me, the most extreme evil is not violence but aggression. Violence is the manifestation of aggression. It is often essential to curb aggression with a big stick, as long as the stick is accompanied by an appealing, convincing idea. Absent such an idea, fanatics of one kind or another will step in to fill the void.

• • •

CONTENDING WITH FANATICISM DOES NOT mean destroying all fanatics, but rather cautiously handling the little fanatic who hides, more or less, inside each of our souls. It also means ridiculing, just a little, our own convictions; being curious; and trying to take a peek, from time to time, not only through our neighbor's window but, more important, at the reality viewed from that window, which will necessarily be different from the one seen through our own.

The fanatic loathes an open-ended situation. Perhaps he does not acknowledge such situations. He always has an urgent need to know what the "bottom line" is, what the inevitable conclusion is, when we will finally "come full circle." Yet history, including the private history of each of us, is usually not a circle but a line: a winding line with retreats and bends, which sometimes changes course and intersects with itself and occasionally draws loops, but nevertheless, a line and not a circle. Being immune to fanaticism entails, among other things, a willingness to exist inside open-ended situations that do not come full circle and cannot be unequivocally settled. A readiness to live with questions and choices whose resolutions hide far beyond the hazy horizon.

WHEN I WAS A CHILD, MY GRANDMOTHER Shlomit explained to me the difference between a Jew and a Christian: "The Christians believe that the Messiah was already here, and that one day he will return. We Jews believe that the Messiah has not yet come but that he will one day." Then Grandma mused: "This disagreement has brought so much hatred and anger to the world, persecution of the Jews, inquisitions, pogroms, mass murders. But why? Why not just agree, all of us, Jews and Christians, to wait patiently and see what happens? If the Messiah comes one day and says, 'I haven't seen you for a long time, I'm so happy to see you again,' the Jews will then have to acknowledge their mistake. But if, when he comes, the Messiah says, 'How do you do? Very pleased to meet you,' then the Christian world will have to apologize to the Jews. Until then," Grandma concluded, "until the coming of the Messiah, why can't we all just live and let live?"

FANATICISM, THEN, BEGINS AT HOME. AND the antidotes to fanaticism may also be found in the home. The poet John Donne gave the world this wondrous line: "No man is an island." To this I dare to add: "No man is an island, but each of us is a peninsula." We are all partially joined to the land that is our family, our language, society, faiths and opinions, state and nationality, while the other side of each of us has its back to all those and its face to the sea, to the mountains, to the timeless elements, secret desires, loneliness, dreams, fears, and death.

THAT SEEMS TO BE THE RIGHT MODE FOR us. The world is full of religions, ideologies, and political movements that propel us to fuse with the collective, to give up being a peninsula, to turn into no more than a tiny speck, a molecule inside the mass of nationality, faith, movement. On the other hand, there are no fewer enormous forces that urge each of us to live like a desert island, to exist at every moment, every day of our lives, in a state of constant Darwinist war against all the other desert islands, because these others, so we are told, are always competitors, always adversaries, and often enemies: if the other has something, you do not have it. If you have it, the other does not.

To be a peninsula is, perhaps, the proper condition for us, despite the fanatics' goal of melting us down completely until we are assimilated in the body of the nation, the faith, or the movement, until we no longer have a shred of privacy, until we are emphatically enlisted in the holy cause. And despite the efforts of other fanatics, who brainwash us that if we are not constantly belligerent, aggressive, and selfish, if we do not take everything by force, we will become weak and lost, and the strong will come and grab everything we own.

Every home, every family, every association, every society and state, every bond between people, including a couple, including parents, is perhaps at its best when it exists as an encounter between peninsulas: close, sometimes extremely close, but without being erased. Without being assimilated. Without revoking one's selfhood.

We all aspire to influence those close to us, to varying degrees, and sometimes we also want to influence those far from us. There is nothing wrong with that, as long as we always remember: influence, without melding. Persuasion, without kneading others into our own mold until they cease being others and become a copy or a satellite of ourselves.

II

Many Lights, Not One Light

In memory of my friends
S. Yizhar and Menachem Brinker

I SHOULD LIKE TO OFFER MY THOUGHTS about Judaism as a culture, rather than just a religion or a nationality. More specifically, I will draw a distinction between things that are no longer relevant and things that are still valid, and furthermore, between rituals and a cherished heritage. There certainly is a Jewish nation, but it is different from many other nations because its lifeblood is not necessarily passed down through genes, nor through victories on the battlefield, but rather through books.

In these times when we are asked to accept that

morality is relative, that what is true for Europe is not true for Africa, what is considered moral in the south is not moral in the east or the west, I sometimes ponder the simple fact that no one in the world is a stranger to pain. Not all pain is equal, yet no normal human being is unaware of causing pain when he or she hurts someone.

Christ told his disciples: "Forgive them, for they know not what they do." I disagree — not with the first part of the pronouncement, because after all, sometimes one can forgive, but with the second part. By claiming that "they know not what they do," Jesus regards all of humanity as morally infantile individuals who commit evil because they do not know that it is evil. In this, he is mistaken and misleading. When we cause pain to another person, or indeed to a cat, we know all too well what we are doing. Even a small child knows. Pain is probably the broadest common denominator in the human race. Who among us has not experienced it?

Pain is a great democratizer. It might even be a bit of a socialist: it does not distinguish between rich and poor, between strong and weak, between renowned and anonymous, between Jew and Gentile, between black and white, between ruler and subject. There are those whose pain has extenuating circumstances and those who are less fortunate, but nevertheless, pain is probably the most comprehensive experience we all share. From this I derive a simple moral imperative: Cause no pain. That is not a sufficient imperative, of course, and we have yet to discuss justice and righteousness, integrity, compassion, pluralism, and so forth. It would be difficult to find two Jews who can agree on which is most important, and it might even be hard to find one Jew who agrees with him- or herself about what came first, what subject has priority, how values should be ranked, and who is authorized to rank them. Some of the Jewish people's fiercest arguments, past and present, stem from disagreements over the question of how to rank values.

It is no accident of history that the Jews do not have a pope. If someone were to stand up and declare himself, or herself, "the Jews' pope," each of us would go up and tap him or her on the shoulder and say, "Hey, Pope, you don't know me, but my grandma and your aunt used to do business together in Minsk, or Casablanca, so please sit down for five minutes — just five — while I explain to you once and for all what God wants us to do."

There is a little guide inside each of us. We are a nation of guides. We all like to teach, to enlighten, to disagree, to shed new light, to oppose, or at least to interpret everything differently. A climate of disagreement is often the right climate for a life of creativity and spiritual renewal. In its good times, the Jewish civilization is one of doubt and disagreement. For thousands of years, Jews added layer upon layer of texts that refer to the texts that preceded them, which in turn refer to even earlier ones. "Refer" does not always mean merely adding on another level or building up another floor. Very often, the new text aims to undermine its predecessors, to show them in a different light, or to suggest a change, an improvement, or a replacement.

The story of Jewish culture is an age-old game

of interpretation, reinterpretation, and counter-interpretation. Not always, though. Not in times of worshiping holy figures, of blind obedience and rote recitation, but in creative times, when Jews incessantly disagree with each other. In such times, Jewish culture is characterized by a vibrant anarchist gene that engenders constant and vehement dispute. How are these disputes settled? "Follow a multitude." That phrase, along with the verse "Beloved is man, for he was created in the image [of God]," represents an ironclad bridge between Judaism and democracy. Follow a multitude — in other words, defer to the majority — not because the majority is always right (in fact it is frequently wrong or iniquitous), but because there is no substitute for the majority's decision, as long as it does not mean oppressing or silencing the minority.*

When a Jewish boy celebrates his bar mitzvah,

---

* The commandment to "follow the multitude" (Exodus 23:2) originally had a narrower interpretation. However, as is customary in Jewish culture, one can and should draw inspiration from the narrow interpretation to reach a broader one.

he is not asked, "What did you learn at school today, my boy?" He is not expected to recite what his teachers told him or what he read in books. On the contrary: "Give us a new idea," he is told. Meaning, tell us something original. Something that is yours. Even a small, marginal, secondary interpretation, but something expressing a thought you yourself came up with after studying the texts. Similarly, a groom on his wedding day was traditionally asked to offer a "novelty" — a new interpretation of a verse or a concept. This seems to be the creative core of Jewish culture through the generations, except in times when it tends to ossify. The Jews did not build pyramids or erect spectacular cathedrals; they did not construct the Great Wall of China or the Taj Mahal. They created texts and read them together in the family, at holiday feasts and at everyday meals.

A LIVELY ARGUMENT RAGES AMONG scholars: How big, or perhaps how small, was Jerusalem, the capital of the Kingdom of David and Solomon? Some say it was "the city of the great king," while others maintain it was just a remote village. One group of scholars even argue that David and Solomon's Jerusalem never really existed; it was a legend. This is a stirring debate, but perhaps it is less important than many of us believe. In Jewish culture and in worldwide consciousness, Jerusalem is not an array of hewn stones, but primarily the city of prophets, of storytellers, of people who formulated notions that changed the foundations of morality. It is the city of Psalms, the city of Ecclesiastes, the city of Song of Songs.

There is an old story that my beloved teacher, the late Mordechai Michaeli, told us when I was a pupil at Tachkemuni, a religious school for boys in Jerusalem. In the story, an elderly father instructs his son: If you seek shelter from the wind and rain, erect yourself a tent or a hut. If you seek a place to dwell in for the rest of your life, build a house of stone. If you wish to take care of your sons and your sons' sons who come after you, build a walled city. But if you want to construct a building for future generations, write a book. That legend might be our calling card: books and family meals. Books and stories that a father and mother read with their children around the holiday dinner table.

In fact, our holidays are all similar: "The bad guys tried to kill us, they couldn't do it, now let's eat." "Pharaoh came, Pharaoh left, bon appétit." On Purim we fought the Persians, on Passover the Egyptians, on Lag b'Omer the Romans, on Independence Day the British and the Arabs, on Tisha b'Av the Babylonians and the Romans, on Chanukah the Greeks. True, on Tu b'Shvat we didn't fight anyone, but on Tu b'Shvat it almost always rains. Of all the destructions and ruinations brought upon us by all those wars, what we have left are the books, the memories, the songs, and the legends.

WHAT IS THE HEART OF JUDAISM? WHAT IS the deepest and most distinct core of Jewish heritage? Perhaps it can be located in its first appearance, on a small potsherd found several years ago at the Khirbet Qeiyafa archeological site, not far from Beit Shemesh. According to the decrypting suggested by Professor Gershon Galil of Haifa

University, it says: "You shall not do it, but worship God. Judge the slave and the widow. Judge
the orphan and the stranger. Plead for the infant,
plead for the poor and the widow. Rehabilitate the
poor at the hands of the king. Protect the poor and
the slave. Support the stranger." Anyone who studied the Tanach as a child, provided they studied it
lovingly and were not brainwashed, will be moved
by this inscription, which preceded the admon-
ishings of the prophets of Israel by centuries. The
inscription's fundamental idea is repeated many
times in the Torah, in the words of the prophets,
and in Jewish heritage. But this Hebrew inscription is likely the most ancient one. More ancient
than the wisdom of ancient Greece. Older than
Rome and all its glory. Fania Oz-Salzberger views
the potsherd as akin to a new text message, sent
from the tenth century BCE to us residents of the
twenty-first century. Written in Hebrew more
than three thousand years ago, the potsherd is in-
scribed with a moral and legal imperative born
from a culture that demanded justice for the weak
and the deprived.

Scholars will undoubtedly continue to argue about whether Khirbet Qeiyafa is the site of the ancient city of Sha'arayim or if it is Neta'im, about whether King David stayed in the palace uncovered in the dig or if it was some other notable personage, whether or not our forefathers ruled the Valley of Elah from that location, and if the site is conclusive archeological proof that the Kingdom of David and Solomon did exist. These are all fascinating questions, but perhaps the crux of the issue is the slave, the widow, the orphan, the stranger, the infant, and the pauper — a meticulous inventory that includes almost every oppressed figure in ancient society. They all managed to squeeze onto this tiny shard, some six inches wide, and to reach us now, of all times. Perhaps in order to show us that a social protest arose in these parts three millennia ago. That the rule of law appeared here before the days of King David. Even before we had a king. There is a huge difference between this Hebrew invention and the Code of Hammurabi and other kings' laws, which they imposed on their ever-obedient subjects. Ancient Hebrew law demands not only the worship of gods and obedience to the king, but is primarily designed to protect the poor, the foreign, the defenseless. "Justice, justice shalt thou pursue" held true even at that time. And further: "Ye shall not respect persons in judgment; ye shall hear the small and the great alike; ye shall not be afraid of the face of any man." Meaning, the law was not intended to magnify the powers of rulers and tycoons.

More than three thousand years ago, there was a culture here that saw fit to demand from the strong that they respect the weak. It demanded not only charity (*tzedaka*) but also justice (*tzedek*) — the two words in Hebrew, unlike in other languages, are closely connected. It demanded this justice not only from rulers, but from every human being.

WHO ARE WE? WHAT ARE WE? THE ANSWER dearest to me is enfolded in that little potsherd. Were someone to suggest that we replace that inscription with a finding more stately, more national, more beneficial to our standing in the conflict about whom this land really belongs to, should we then trade that little piece of pottery for a graffitied inscription on the remains of a wall, in handwriting verified as David's, reading, "I, King David, son of Yishai, slept here"?

No. We should not.

OTHER JEWS LOCATE THE CORE OF JUDA-ism in different places. Some find it in the 613 commandments, in the prayers, in erudition, in the tombs of holy men, in the secrets of Kabbala and various signs and symbols and miracles, or on the Temple Mount and the renewal of the sacrificial rite. All of these approaches are well founded on ancient texts. Some of them are dear to my heart, others do not speak to me, and yet others I find abhorrent, just as there are Jews who find my own attitude abhorrent, in part because I refuse to revere rabbis, holy men, teachers, and their disciples who never doubt that they know God's will absolutely.

Again: it is no coincidence that we do not have

and could never have a pope. Extraordinary rabbis and scholars have always existed and still do, but there is almost always more than one, and there is almost always disagreement among them.

We conduct passionate arguments about democracy. What is democracy? What is so good about it? Does it come at the expense of Judaism, or is it the other way around? How can we reconcile democracy with Jewish heritage? Modern democracy stems from humanism. There is no contradiction between Judaism and humanism. In *The Fathers According to Rabbi Nathan* (a collection of commentaries on *Ethics of the Fathers*, probably composed in the sixth century and included in the Talmud's extra-canonical tractates), chapter 31, Rabbi Nehemiah formulates the very heart of the humanistic idea in nine wonderful words: "One person is equal to the whole of creation." (Note that he says "one person" — he definitely does not say "one Jewish person." But Judaism has featured very different voices, too, including ones full of arrogance and xenophobia.) Humanism is also bound up with pluralism — namely, acknowledging that human beings have a right to be different from one another and that each individual being is akin to a whole world worthy of a dignified existence. In much the same way that I sometimes reduce all the commandments into one — Cause no pain — I am sometimes willing to narrow down humanism and pluralism into one simple formula: Recognize the equal right of all human beings to be different.

There are those who recite to us, day and night, that "united we stand." Indeed, our strength is in being united around our right to be different from each other. Being different is not a passing affliction but a blessing. Disagreement is not a troubling state of weakness, but a vital climate for the growth of a creative life. We are different from each other not because some of us have yet to see the light, but because the world is full of many lights, not one single light. Many beliefs and opinions, rather than one belief and one opinion.

We have been taught since childhood that the earlier iterations of the State of Israel were brought down by internal strife, having collapsed under the weight of "unjustified hatred." Lately there are those who insist that if we finally set aside our disagreements and unite as one, we will vanquish the whole world. But the truth is that the uprising against the Romans, which led to the destruction of the Second Temple, just like the earlier war against Babylonia, which had caused the destruction of the First Temple, did not fail because of "brotherly strife" or "unjustified hatred" among Jews, but because of nationalist and religious fanaticism, because of the delusions of grandeur suffered by both leaders and subjects who had lost all sense of reality. Even if the Jews of the First and Second Temple eras had loved each other more than David and Jonathan did, even if they had banded together as tightly as a block of concrete or steel, Babylonia and Rome would still have effortlessly crushed the impudent little nation bent on slamming its head against the wall.

The destruction of both temples did not stem from "unjustified hatred"; it was the fault of the zealots who lost all sense of proportion and reality and dragged the people of Israel into a fatal clash with forces infinitely stronger than they were. They had unwavering and delusional faith that God would intervene at the last moment and drown Pharaoh and his chariots. Today's Israel faces similar dangers if our current fanatics continue to lead us into conflict.

THE JEWS AS A PEOPLE ARE NOT DISPOSED to obedience. Never have been. Moses could tell you how unaccustomed the Israelites are to being obedient. The prophets could also attest to it. God Himself complains throughout the Bible that the Israelites are insubordinate: the people argue with Moses; Moses argues with God; Moses hands in his resignation and eventually backtracks, but only after negotiating with God, who gives in and accepts his main demands (Exodus 32:31–33). Abraham bargains with God over Sodom like a used-car salesman: Fifty righteous men? Forty? Thirty? Twenty? Would you settle for ten? And when it turns out that there are not even ten righteous men in Sodom, Abraham does not fall to his knees and beg God to forgive him for his impudence. On the contrary. He looks up to the heavens and utters what might be the boldest words in the Bible, if not in all the history of religion: "Shall not the Judge of all the earth do justly?" (Genesis 18:25) In other words: You may be the judge of this entire land, but you are not above the law. You are the legislator, but you are not above the law. You are Master of the Universe, but you are not above the law. You are the Creator of heaven and earth, but you are not above the law. We find no such talk in Christianity, in Islam, or in any other religion I know of. And it is a quality worthy of praise.

Moreover, Abraham is not struck by lightning as punishment for his impertinent offense of God. However, a few chapters later, the same Abraham is willing to sacrifice his son, Isaac, out of blind obedience. How might we bridge the chasm between the Abraham who fights God on behalf of strangers, the inhabitants of Sodom, and the Abraham who does not hesitate for a moment when God commands him to slaughter his own son? The Israeli writer Shulamith Hareven offers a fascinating interpretation of the Binding of Isaac. Like all commentators, Hareven agrees that Abraham was being tested. But contrary to the traditional readings, she believes that Abraham utterly failed the test, because in fact he should have refused the order. He should have resisted the command and told God, "You yourself forbade us to make human sacrifices, and so I refuse to sacrifice my son." God put Abraham to the test, and Abraham, the glorified "knight of faith," failed because he said, "Yes, sir!" instead of declaring, "That is a patently illegal order with a black flag waving above it."*

The nation fights incessantly with the prophets; the prophets fight with God while also squabbling with the people and the kings. Job hurls his anger heavenward. God refuses to admit that He has wronged Job, but nevertheless awards him personal compensation. The Gemara, too, contains stories that express a measure of rebellion

---

* Surprising support of Hareven's audacious reading can be found in the Babylonian Talmud, Tractate Ta'anit 3:71. Also see Jeremiah 19:5.

against God, and even a reluctance to accept God's law. The most turbulent of these Talmudic stories is a tale known as "The Oven of Akhnai," in which Rabbi Yehoshua insists that "it is not in heaven" when the Lord purports to be the decisive voice in an argument between the sages.

And in recent generations Hasidic rabbis have demanded that God appear before a rabbinical court to justify the terrible things occurring in this world. God, of course, does not appear before the rabbis, but the summons instructing Him to be present before the rabbinical jurisdiction is still valid. The rabbis wait in the earthly court for Him to come and justify Himself, to give meaning to the suffering and injustice, to finally explain why "the righteous suffer and the wicked prosper."

This rebellious streak is interwoven throughout Jewish history: "Shall not the Judge of all the earth do justly?" "How long shall the wicked, Lord, how long shall the wicked exult?" A thread leads from the offenses hurled at the heavens in the Torah, in the Prophets, in the book of Job, in the Gemara, and in Hasidic tales, all the way to Uri Zvi

Greenberg's wonderful yet horrifying poem, "At the End of Roads Stands Rabbi Levi Yitzhak of Berdichev and Demands a Great Answer." In this post-Holocaust poem, Rabbi Levi Yitzhak turns to God and castigates Him by asking, more or less, "Where were you? How could you?"

Decades later, in his poem "God Full of Mercy," Yehuda Amichai wrote, "Were God not full of mercy / there would be mercy in the world, and not just in Him."* It is Amichai rather than the holy man of this place or the *gaon* of that, Amichai and not rabbis or experts in Jewish law, who to my mind is the true heir to the finest of Jewish culture. In this poem and in others, he carries the psychosocial and moral genes of Abraham arguing for Sodom, of the prophet Jeremiah, of the book of Job, of the Oven of Akhnai, of Rabbi Levi Yitzhak of Berdichev. Even in the writings of S. Y. Agnon, a devout Jew, we find characters who lambaste God and make horrifying accusations. Daniel Bach, the protagonist of *A Guest for the Night*, says, "I

---

* Translated by Robert Alter.

am a simple man, and I do not believe that the Almighty wants the welfare of His creatures." True, Bach is far from saying that there is no God. He is not disputing that God is almighty. Daniel Bach certainly believes in God, but he is full of fury at Him. He does not believe that God wants the best for His creatures.

This is the anarchist core, the rebellious gene that has flickered for thousands of years in Jewish culture. We don't want simple discipline. We don't just follow orders. We want justice, and we demand it even from the Creator. "Justice, justice shalt thou pursue." Or, as Bialik phrased it in "On the Slaughter": "If there's justice — let it come now! / But if it should come after I've been / blotted out beneath the sky, / let its throne be cast down."* (Again: there have been and still are different Jewish voices, which yearn for victory and vengeance and not for justice and compassion.)

A young man searching for lost she-asses and

---

* From "On the Slaughter," Chaim Nachman Bialik, translated by Peter Cole, *Paris Review*, 2014, https://www.theparisreview.org/blog/author/pcole/.

a shepherd struck by inspiration can be anointed king of Israel or write the book of Psalms. A sycamore fig farmer can turn into a prophet. An illiterate shepherd, an unknown shoemaker, a work-weary blacksmith, or even a rehabilitated thief can go on to teach the Torah, write interpretations, and leave their imprint on the daily lives of every Jewish person for millennia. Yet they were each dogged, almost always, by the question, "Who made thee?" In other words: How do we know you are genuine? You may really be a sage of the Torah, but on the neighboring street lives another sage, who proposes a completely different conclusion from yours. Not infrequently, it transpires that "both these and those are the words of the living God."* And often the disagreement is not a curse but a blessing: "To make the teaching great and glorious."† "An argument for the sake of heaven's name."‡ "Jealousy among teachers increases

---

* Talmud, Eruvin 13:b.
† Isaiah 42:21.
‡ Mishnah, Pirkei Avot 5:17.

wisdom."* And in a rare moment of grace, the Lord Himself might admit His error, smile, and say, "My children have triumphed over me."†

In Jewish history, the question of interpretive authority has usually been decided by partial consensus rather than unanimously. The history of Jewish culture in the past millennia is a series of bitter disputes, including a few tinged with malicious urges, and some wonderful and fertile ones. The Jewish people often had no authoritative mechanism for formal decision-making, at least not since the decline of the Sanhedrin. We did not have a system in which a man in a white cape overrides those who wear red capes, while the red-cape wearers prevail over those who wear black capes. A particular rabbi might be considered greater than another rabbi simply because many people think him greater.

Jewish culture has been built, generation upon generation, with the creative energy arising from tension between the Kohen and the prophet, be-

* Mishnah, Bava Batra 21:a.
† From the aforementioned Talmudic tale "The Oven of Akhnai."

tween Pharisees and Sadducees, between the House of Hillel and the House of Shammai, between Sephardic and Ashkenazi prayer services, between Hasidim and their opponents (Misnagdim), between the devout and the proponents of Jewish Enlightenment (Haskalah), between Zionists and anti-Zionists, between the Bialik school and the Berdyczewski school of poetry, between religious and secular, between hawks and doves — to this very day.

Jewish culture at its finest is a culture of give and take. Of negotiating. Of cutting both ways. Of acuity and powers of persuasion. Of "both these and those are the words of the living God." And sometimes of ties: when the greatest minds were unable to agree on a solution, they would declare a tie, or in Hebrew, *teko*, a term still used in colloquial Hebrew, and which, according to one derivation, is the acronym of an Aramaic phrase meaning "the Tishbite [Elijah] will settle the question when he comes." In other words: never mind, we'll agree to disagree until Elijah the prophet turns up and makes the call. It's no terrible thing to disagree. We can certainly live in an open-ended situation. It may even have its advantages. But there is one crucial caveat: arguments must be conducted without resort to violence. Disagreements without persecution. As Prime Minister Menachem Begin is said to have instructed the head of the Shin Bet when interrogating detainees: "Not even a slap on the cheek."

Jewish culture sanctifies disagreement for the sake of heaven's name. It encourages dissent. Undeniably, it is also sometimes a culture of aggressive impulses disguised as disagreement for the sake of heaven's name—impulses of power, authority, and honor. And to my mind, this heritage coexists superbly with the ideas of pluralistic democracy. One might say that the tradition of arguments and disputes in Jewish culture is analogous to the musical concept of counterpoint, as well as to the notion of human polyphony, whereby the community is viewed as a chorus of different voices, or different instruments orchestrated by an agreed set of rules.

. . .

MANY LIGHTS, NOT ONE LIGHT. MANY
beliefs and opinions, not one.

JEWISH CULTURE HAS FREQUENTLY HAD
large pockets of blind obedience among those who
see Judaism as purely a religion and not a culture.
I regard blind obedience as a deviation from tradition even when it purports to be the embodiment
of that tradition. Despite the enormous differences between the domineering Litvak rabbis and
the Lubavitcher "Messiah," or between the former
and the Baba Sali (Rabbi Israel Abuhatzeira, a leg-
endary rabbi and Kabbalist of Moroccan descent),
or between all of these and the rabbi who founded
Shas, the Orthodox Sephardic political party, they
all share an aspiration to impose popish obedience
on all who surround them. Their subjects submis-
sively accept the yoke of discipline.

To me, blind obedience can never be moral. In
this I am far removed from the Judaism of Yeshayahu Leibowitz, an iconoclastic philosopher
and scientist who, despite being a thorn in the
side of the Israeli right wing, was devoutly Or-

thodox and held that the biblical commandments represent God's will and must not be questioned or deviated from.

IN POSTBIBLICAL TIMES, THERE HAS NOT been a single event regarded by all Jews as a miracle. Many people believe in miracles these days, but the dubious outnumber them. Almost every figure of authority soon gives rise to a counter-authority; at least this is how things go in good times. There have been only a few individuals in Jewish history whose authority was accepted unquestioningly by both their own peers and successive generations. Ultimately, the source of authority in Jewish culture is the willingness of the nation, or much of it, to accept the person who issues an instruction or makes a Halachic ruling, or the miracle-making *tzaddik* or the guide, as a source of authority. (Halacha is the body of Jewish laws derived from the Bible and from its rabbinical interpretations.) Even Maimonides, who is referred to in our sources as "the great eagle," became great because the people took him into their hearts, not because he was crowned by a handful of cardinals who sat around debating until a white smoke signal emerged.

The hierarchy of authority among the Jewish people, at least in good times, is not determined from above. In this respect, again, there is a profound democratic streak in Jewish culture through the ages. This is worth emphasizing at a time when prominent figures are not content to grapple only with the genuine contradiction between the government's authority and that of Halacha, but insist also on confronting democracy with Judaism, defining the spirit of democracy as a threat to the spirit of Judaism, or Judaism as a threat to the principles of democracy.

The philosopher Isaiah Berlin teaches us that the controversial question among democrats is whether political liberty is essentially negative, as in "live and let live," or positive, as in "live correctly in order to be truly free."

FROM MY DAUGHTER FANIA OZ-SALZberger, I learned that liberal democracy is a way for a society or a state to organize itself, in which the explicit purpose is to reconcile the desires of all the individuals who belong to it, while preserving their freedoms. It negotiates these different desires by means of elections and majority decisions, while maintaining the rights of every minority through a system of compromises and protections. I further learned from her that the most ardent democrats of early modern history were, surprisingly, religious extremists, such as the Huguenots in France and the Levellers in England, who fought against the regime's attempts to impose the majority religion on them (although they did not act with particular tolerance among themselves).

THE WORLD OF HALACHA, LIKE THE UNIverse itself, began with a Big Bang, when Moses was given the Torah on Mount Sinai. Ever since, and until only recently, Jewish culture has been one of ever-expanding ripples, as though a giant meteor fell into an ocean and the effects are still rippling out in ever wider circles around the revelation at Sinai. Expanding waves of interpretations, and interpretations of interpretations. But as these ripples expand, they naturally also weaken. Jewish heritage over the generations is made up of layer upon layer of interpretations of the Torah. The farther away one gets from Mount Sinai, the narrower the interpretive realm becomes, because it is more and more populated. Each generation populates the interpretive space ever more densely, and no generation is allowed to detract. No one is allowed to do away with anything. The result is that the house fills up with furniture, the furniture gets piled up with objects, no one departs, and before long no one will be able to enter the space. In the world of Halacha, scholarly expertise, meticulous devotion, acuity, and enthusiasm are on the rise as the creative expanse narrows.

Scholars' sense of self-worth also shrinks from generation to generation, because the convention holds that "if the earlier [scholars] were sons of angels, we are sons of men; and if the earlier [scholars] were sons of men, we are like asses,"*

---

* Talmud, Tractate Shabbat 112:b.

and there is an overall perception of "the decline of the generations." As we grow further away from Mount Sinai and the giving of the Torah, Halachic Judaism preserves and declaims more and more, while creating less and less. And newcomers are not permitted to alter their predecessors' words.

While it is true that spiritual clocks showed different times in the different Jewish communities of Baghdad, Yemen, Morocco, Salonika, and Eastern Europe, all the clocks now show that Mount Sinai is increasingly enshrouded in a pillar of smoke made up of layers of interpretive text. This is one cause of the smothered feeling that many people experience around the world of Halacha.

Up to a certain point, the Judaism of *Shulchan Aruch* (literally "The Set Table," a sixteenth-century collection written by Joseph Caro and widely considered the principal codification of Jewish law) withstood the pressures and temptations of external worlds with surprising success. This was possible not only because of the Jews' religious devotion, but also because of the similarities between their lives in Eastern European or North

African enclaves and the lives of their non-Jewish neighbors, who adhered to their respective religions and lived in pious communities centered around the church or the mosque. As long as the Jewish "reservations" maintained a separate but similar existence alongside their Christian or Muslim neighbors, with a marked religious focal point, it was possible to preserve and protect their distinct identity. But the more open, tolerant, and curious the non-Jewish surroundings became — for example, in the golden age of Muslim Spain — the more the Jews themselves were drawn to a less stringently Halachic lifestyle. The creative horizon expanded.

When secularism began to spread in Europe some two centuries ago, and the non-Jewish environment's identity became less focused on religion and more on nationality, and even more forcefully since the appearance of multinational or supranational secular ideologies, the Jews' lives within the walls of Halacha became all the more burdensome, while the charms of nearby worlds seemed increasingly alluring. (S. Yizhar addresses the issue in his stirring essay "The Courage to Be Secular.") This development occurred in part because the affinity to the Other is an essential component of every individual's and group's identity. Indeed, our relationship with our enemies is part of our self-definition. As affinities to Others became more fluid and seductive, as the identities of those around us changed, our own identity was inevitably rattled as well.

True, millions of Jews who call themselves God-fearing stood fast in their faith, even amid widespread tremors within Halachic Judaism: the Sabatteans, the Hasidics, the appearance of huge yeshivas in the nineteenth century, the birth of religious Zionism. In recent generations, increasing numbers of Jews were no longer satisfied with living strictly according to Halacha. Some expanded their identities in national political directions, others toward religious reform. A great many headed to the exit door and assimilated. Halachic Judaism responded with panic and fury: shunning, excommunicating, cursing, swearing, toughening up, and hunkering down, as if it had decided that all God-fearing Jews must buttress themselves inside a spiritual and emotional bunker until order was restored. The majority of Halachic Judaism has yet to leave that bunker, including in Israel, in the ultra-Orthodox neighborhoods of Jerusalem and in predominantly ultra-Orthodox cities like Bnei Brak, Beit Shemesh, and Beitar Illit.

Nationalism, emancipation, integration within the local culture, cosmopolitanism in all its varieties, modern existence, doors that were opened to Jews in several countries in the eighteenth and nineteenth centuries — Halachic Judaism responded to all these reversals as though they were passing aberrations. Not only did the religious establishment resist adapting to the new circumstances, but it refused to examine them or so much as glance at them. In fact it refused to admit that there *were* new circumstances. To this day, Halachic Judaism insists that although something may look new, it is nothing but a camouflaged reincarnation of old temptations: "It's all happened before." "The new is forbidden by the Torah."* As if the prohibition against any innovation not only applies to Jews but must also extend to Gentiles, to the external world. Even the murder of millions of Jews by the Nazis is still depicted with pathetic clichés by the hardened sectors of Halachic Judaism. Over and over they recite: "In every generation there rises up a Pharaoh, an Amalekite, a wicked Haman, an Antiochus." As though the Holocaust were simply another pogrom, albeit a worse one. Here come the Cossacks again, and the Inquisition, and the anti-Semites; here comes Amalek, and Pharaoh. As though the Nazis were just another desert tribe attacking the stragglers. As though a genocide of the Jewish people were merely a link in a long-familiar chain, the chain of *tsuris* — of grief. Yet another holy sacrifice, another test we were subjected to "because of our sins," a test whose outcome can be reversed by repenting.

---

* Talmud, Kiddushin 38:b.

This is how *Shulchan Aruch* Judaism evaded and still evades a genuine theological reckoning with the murder of a third of the Jewish people. On this matter there have been dissenting voices here and there in the ultra-Orthodox world, such as that of Rabbi Yosef Shlomo Kahaneman, founder of Ponevezh Yeshiva in Bnei Brak.

Non-Zionist Halachic Judaism has similarly fled from contending, theologically, with the unprecedented phenomenon of the Jewish people's political independence in the Land of Israel, with the construction of Jerusalem — not by an angel or a seraph, not by the coming of the Messiah, but rather by a secular, modern political movement, one very much influenced by the national movements that arose in other parts of the world.

IN AN UNTHINKING TURN OF PHRASE, THE Nazis' victims and the casualties of Israeli-Arab wars were all perceived to have died for "the sanctification of God's name." But the millions of Jews murdered by the Nazis and their accomplices were not killed to sanctify God's name. It infuriates me to hear the phrase "may God avenge their blood," since millions of them did not believe in sanctifying God's name in the first place. Hundreds of thousands of the victims, who were born to non-Jewish mothers, were not even considered Jews by *Shulchan Aruch* Judaism, so what gives them the right to attribute "the sanctification of God's name" to these deaths? What gives Halachic Judaism the right to appropriate their identities and trample their dignity? Very many of the casualties would probably have considered the label of "martyr" or "killed for the sanctification of God's name" a crude desecration of their memory, of their identity, of their self-determination.

Most of the Jews killed in wars between Israel and Arab states did not enlist in order to sanctify God's name either. They went into battle to protect their lives, their loved ones, and their nation. Many hundreds of those killed in Israel's wars were not Jews at all: they included Muslims, Christians, Druze, Bedouin, Circassians, and people of other nations who came to Israel as volunteers and gave their lives, most certainly not to sanctify God's name. Not at all.

In the early 1950s, the Israeli passengers of a bus were murdered at Ma'ale Akrabim in the Negev. The minister of defense declared on the Knesset podium: "Jewish blood shall not be forsaken!" Here we must pause and emphasize that this horrible phrase, "Jewish blood," does not appear anywhere in Jewish sources. Not once. We do not have "Jewish blood" as a concept. In the Bible there is "clean blood." There is "the voice of thy brother's blood crieth unto Me from the ground." There is "Whoso sheddeth man's blood, by man shall his blood be shed." Later, we have the saying "Your blood is no redder," meaning, no redder than anyone else's blood, for we were all created in the image of God. Conversely, "Jewish blood" was a central concept in the Nuremberg Laws legislated by Hitler's Reich. Recently, this monstrous concept has become popular among many of the more extreme Israeli Jews.

A few years ago, my wife Nily and I were guests

at the home of an old friend who is no longer with us. He was a great intellectual whose opinions were aligned with those of the Greater Israel movement. Nily, a talented musician, sang us a well-known Shabbat song based on a verse from Psalms: "When the Lord brought back those that returned to Zion, we were like unto them that dream." Our host's eyes filled with tears, and he made a heartfelt plea: "We should cancel 'Hatikvah,' and instead our national anthem should be 'When the Lord brought back those that returned to Zion, we were like unto them that dream.'" I replied: "Absolutely not! If we're going to change the Zionist anthem, it should be 'We have not experienced a miracle, we have found no jug of oil,' which is the complete antithesis of 'When the Lord brought back those that returned to Zion.'"

Those who would blur or diminish the fact that the modern "return to Zion," the construction of Jewish towns and kibbutzim and cities, did not derive from the Messiah but in fact from a secular, pragmatic, modern political movement, threaten the Jewish identity cherished by me and others like me. They threaten to erase us. Not to mention the deeply insulting conviction of some of Rabbi Kook's disciples who maintain that the secular pioneers were no more than unwitting instruments of divine supervision, and that everything they believed in, their entire self-determination, made no difference. They were nothing more than "the Messiah's donkey." Such an insult is intolerable.

SURPRISINGLY, IT IS IN THE WORKS OF modern Hebrew authors — Judah Leib Gordon and Micha Josef Berdyczewski, Bialik and Brenner and Agnon, Uri Zvi Greenberg, Zelda, S. Yizhar, Pinchas Sadeh, Yona Wallach, Yitzhak Orpaz, Yehuda Amichai and Dan Pagis, Haim Be'er, Haim Sabato, and Zeruya Shalev — where we find many profoundly religious moments. Several Hebrew writers in the past century or so have, ironically, assumed the role from which various Halachic authorities have fled.

We are dealing with a separation of religion — no, not from the state, but from religious people: the most important theological events in Jewish culture over the past few generations did not occur in seminaries or rabbinical courts, but in Hebrew poetry, prose, and philosophy. It is exciting to recognize that theological grappling has not vanished from our culture, but rather has been passed from the old gatekeepers to Judaism's most creative force: new Hebrew literature. Its creators have never let God alone. They have insisted on tugging at His sleeve, trying to decipher His intentions, to prosecute Him, badger Him, sometimes pour out their wrath at Him, and sometimes to yearn for Him profoundly. Writers who essentially perceive themselves as secular have nevertheless persistently expressed theological distress. For example, at the end of S. Yizhar's *Khirbet Khizeh,* after the titular Arab village is destroyed, "then God would come forth and descend to roam the valley, and see whether all was according to the cry that had reached him."* Many writers, poets, and intellectuals have written, albeit without naming the term explicitly, about the concept of *hester panim,*

---

* Translated by Nicholas de Lange and Yaakob Dweck.

which literally means "the hiding of the face," and connotes the apparent absence of divine intervention in times of calamity. These are the living wellsprings of Jewish culture in our generation. These are the continual fountains of knowledge in our age. These, and not the yeshivas.

Halachic Judaism has become a repository that, although it does not lose a single drop, adds virtually none either. Most of the dynamic, creative revelations in Jewish culture have occurred in recent generations outside the world of Halacha, even if they have a dialectical affiliation with it. At times they may fulminate against it, but that, too, is an affiliation. Sometimes a critical relationship is more intimate than the kind maintained by those "curators" who place all of Jewish heritage behind fortified glass and slap a Do Not Touch sign on it, hoping that future generations will merely stand before the glass, behold the glory of their legacy, and then memorize, internalize, and pass it on.

Halachic Judaism often claims that the Torah is what protected the Jewish people, and that without it we would have long ago assimilated among the nations of the world. But the truth is different: it was not the commandments that protected the Jews, but the Jews who endeavored to preserve the commandments, by observing them or in some other way choosing to be Jewish. The people of Israel have existed for thousands of years solely because of millions of personal decisions made by millions of Jews for dozens of generations — Jews who chose to maintain their identity.

The Torah and the commandments, the languages that Jews spoke, their collective memory, their lifestyles and sensibilities and creative acts — all these always existed, thanks to a personal, private decision made by a Jewish individual: to stay and not to leave. Identity is meaningful only when the exit door is wide open. Only when permission is granted. Only when each individual chooses, of his or her own accord, to maintain that identity and not replace it.

Agnosticism is also a part of Jewish culture. So is heresy, so is excoriating God — these are all in fact markedly religious positions. The writer Shlomo Zemach once observed, insightfully, that a person does not curse God if there is no God in his heart.

Consider, for example, Elisha ben Abuya, a first-century Jerusalem rabbi who rode his horse on the Sabbath, among other transgressions, and was referred to as "the Other One." He was denounced, he was excommunicated, yet his writings are included in the Talmud. According to one Talmudic version, he was even destined to reach the afterworld. Perhaps the day will come when the excommunicated Spinoza, or Heinrich Heine — a convert who nevertheless remained Jewish to the bone — or even that tortured Jew, that wonderful poet and paragon of morality from Nazareth — perhaps all of them will one day regain their prominence in the Jewish pantheon. They are all flesh of our flesh.

What does Jewish culture comprise? It comprises everything we have amassed over the generations. Elements born inside it, as well as those absorbed from the outside, which become part of the family. Things that are customary, or used to be; things we all accept, as well as things only a few accept. Things accepted today, and things accepted in previous generations. Aspects that I accept, and ones I find annoying and distasteful. They are all included in Jewish culture. Texts written in Hebrew, and texts written in other languages. The written word, and ideas found outside of books, including certain behavioral inclinations and responses connected to our shared memory. A certain type of humor and a tendency to wisecrack, which I cannot define but which I easily recognize whenever I encounter them. A blatant inclination to be critical and skeptical, to be ironic, self-pitying, and sometimes self-righteous. A particular pragmatism tinged with fantasy. Ecstasy diluted with doubt. Euphoria blended with pessimism. Melancholy cheerfulness. And a profound, healthy suspicion of authority. A measure of stubborn resistance to injustice.

These contradictory and complicated features would seem to characterize us. But they are not present in every individual. Moreover, there is no assurance that they will continue to exist in the future. Jewish sensibilities can be identified without much difficulty in Jesus and Heine, in Spinoza and Einstein, in Ibn Gabirol and Kafka, in Karl Marx and the Marx Brothers, in Hannah Arendt and Woody Allen, in the prophet Jeremiah and S. Yizhar, in the author of the Psalms, in Ecclesiastes, in Rabbi David Buzaglo, in Zelda and Yehuda Amichai. It is not hard to spot these sensibilities, but it is almost impossible to tie them down with definitions. Quite a few of them are slowly vanishing in Israel, along with the imperative to "cause no pain."

WHEN IT COMES TO POLITICAL ATTITUDES, there is, paradoxically, an alignment between two opposing extremities of Halachic Judaism: on the one hand the anti-Zionist ultra-Orthodox (known as Haredim), and on the other the Messianic Jews of the hilltop youth settlements, who are in fact post-Zionists. Both find it extremely difficult to accept the State of Israel as it is, with an elected government, with non-Jewish citizens who are supposed to have equal rights, with the control tower of the Supreme Court. The Haredim are unwilling to accept the Knesset and the Supreme Court as substitutes for the Sanhedrin, or to recognize Israeli law and Supreme Court rulings as superseding Halachic adjudicators. They also deny the authority of the State of Israel, despite the Halachic rule whereby "the law of the land is the law," which ensured that their forebears conceded to the authorities of the Polish or Russian governments, or to the king of Morocco. The Haredim are incapable of accepting the State of Israel as either a binding Jewish authority or a binding foreign one; as the Hebrew saying goes, they can neither swallow nor vomit.

The Messianics have a different problem with the State of Israel. Many of them view the state as merely a shell or a scaffolding, and now that they can hear the Messiah's footsteps, they are eager to knock down the scaffolding and heroically produce the kernel of monarchy from within the shell. They repeatedly tell us that democracy is a foreign element, an import. We do not need it. We are tasked with rebuilding the kingdom. My friend Avi Ravitsky justifiably reminds these Messianic

Jews that while the issue of whether or not democracy is foreign to us might be debatable, there is no question that sovereign rule was a foreign import. It was not the Jews who invented monarchy. The masses imposed it almost forcibly on the prophet Samuel. Before the Israelite kingdom was established, during the days of Judges, the regime resembled a republic more than a monarchy. Monarchy is a distinctly foreign type of rule, far more alien to Jewish heritage than democracy.

The ghetto is not a Jewish creation either, nor is it in our roots. It was imposed on us by foreign decree, and some of us fell in love with it and still try to maintain it in Israel, because only inside a ghetto do these Jews feel warm and safe.

Shulchan Aruch Judaism is virtually unable to sustain political life: without the authority of democracy—that "foreign element" instituted by the secularists—the various factions of Halachic Judaism would squabble among themselves, and "one person would eat the other alive."* Without

---

* Mishnah, Seder Nezikin, Tractate Avot 3:2.

the ingrained tradition of making decisions by majority, it would be Hasidim against Misnagdim, Hasidim against other Hasidim, one court against another court, Sephardim against Ashkenazim, and all the various gradations of religious adherents against one another. If, God forbid, the State of Israel were to disappear overnight, along with its parliament, elections, and courts, all the streams of Halachic Judaism would immediately need a foreign regime to adjudicate. They would be running to the Arab government day and night to come and settle their disputes, just as they did in the Diaspora. When future president Chaim Weizmann was asked when there would finally be a Jewish state, he astonished his audience by replying, "Never." Then he explained, with a smile, "If it is a state, it shall not be Jewish. And if it is Jewish — it shall not be a state."

Shulchan Aruch Judaism has no mechanisms for governing, either by majority rule or through elected institutions. There is only a long tradition of competition among rival religious groups for the graces of the ruling foreign regime, and of flattery of its representatives. Members of each religious faction are convinced that *theirs* is the true Judaism and that anyone who is not like them is deviant or sinful or deaf.

Almost none of the many contemporary streams of Halachic Judaism are capable of lovingly resuming the Jewish pluralism of the better eras. Almost none of them can experience the wonderful gift embodied in the saying "Both these and those are the words of the living God." (Exceptions include the egalitarian Orthodox community Shira Hadasha and the founders of the poetry journal *Mashiv Haruach*.)

Halachic law itself is practically in the Stone Age when it comes to political issues. Its sophistication regarding questions of money, family, parenting, education, interpersonal relations, community, and property does not extend to relations between nations, or between Jews and other peoples, where it seems to recognize only one of two conditions: "the Gentiles have the upper hand" or "the Jews have the upper hand." When it is the former, the Gentiles abuse us and we cry to the heavens for mercy. When the latter — why should we not abuse them a little and have them cry out? That is more or less what you will find, for example, in the book *The King's Torah*, which the authors dedicate to the memory of Baruch Goldstein, the Jewish mass murderer who killed twenty-nine Palestinians at the Tomb of the Patriarchs.

The Jewish people have simply not had the opportunity nor the suitable life experience to enable them to develop sophisticated laws regarding relations between Israel and other states. Indeed, there are observant Jews who are willing to be endlessly welcoming and to shower those around them with love, but it is always a conditional love: my love for you, they say, depends on you changing and becoming at least a little bit like me, but I will not change one iota, because I am right. I am perfect.

Psalms 81:11 says: "Open thy mouth wide and I will fill it." Various entities that try to convert secular Jews to religious observance, such as the television channel Hidabroot and the like, the Lubavitchers, some of the settlers, the Misnagdim, and the Hasidim, all love us conditionally. The condition is: you open thy mouth and I will fill it. Because you, the Other, are an empty vessel, while I am a vessel filled with blessings. This condescending commandment, "Open thy mouth wide and I will fill it," contains a clear undertone of violence, which is, of course, the complete opposite of dialogue.

David Ben-Gurion once visited Rabbi Avraham Yeshayahu Kerlitz, a prominent figure known as the Hazon Ish, at his home in Bnei Brak shortly after the founding of the state. The Hazon Ish offered Ben-Gurion a biblical parable: A full cart and an empty cart approach each other on a narrow bridge. Is it not just that the empty cart make way for the full cart? Ben-Gurion, perhaps in a moment of intellectual weakness, took upon himself — and upon us — this law of the empty cart. But the Jewish culture that evolved outside the synagogue over the past few centuries is no empty cart. Bialik and Agnon, Brenner and Berdyczewski, Rachel the Poetess, Uri Zvi Greenberg, Nathan Alterman and Leah Goldberg, Gershom

Scholem and Martin Buber, S. Yizhar and Yehuda Amichai all tow carts of Jewish heritage no less full than those of the great religious adjudicators. The revival of Hebrew, and its transformation within three or four generations from a sleeping beauty into the daily spoken language of more than ten million people, is a spiritual occurrence no less vital to Jewish history than the emergence of the Talmud. Cities and towns, villages, kibbutzim and moshavim that were built in Israel in the past 120 years are, to me, among the most fascinating creations of the Jewish people in history.

A spiritual dialogue between a full cart and an empty cart is not possible. Only those willing to acknowledge that Jewish culture is made of several full carts can be true interlocutors.

AT LEAST SINCE THE FOUNDING OF IS-rael, the religious parties have employed politics in order to strengthen what they absurdly call "the Jewish character of the state," which is a term that currently has no meaning. At most, one might speak of "Jewish characters" — in the plural.

The character of Tel Aviv is no less Jewish than that of Bnei Brak. The character of development towns like Sderot or Yerucham, or of the kibbutzim, is just as Jewish as that of the historic city of Safed. Anyone who believes that "Jewish character" can mean only a replica of the Jewish shtetl in Poland, or a continuation of the Jewish *mellah* in Morocco, or a reconstruction of the Kingdom of David and Solomon, is shutting himself inside a recalcitrant bubble. Indeed, it would not be a bad thing for the State of Israel to bear certain similarities to any of the aforementioned examples, but it would also be a good idea if it could resemble, just here and there, other Mediterranean locales.

HAS ISRAEL BEEN MOVING FURTHER AWAY from my own ideal of a Jewish state in recent years? Certainly it has. And the imperative to "cause no pain" is also fading.

The State of Israel is the child of a mixed marriage. It was born from a merger of the Bible with the Enlightenment, of the longing to return to

Zion with the sentiments of the Spring of Nations in Europe, of the Jewish communities with the republic, of "Knesset Israel" (which originally meant "the nation of Jews") with the parliamentary spirit, of modes of life that evolved in the Diaspora's villages with modern ideologies that champion liberty and fraternity among all people.

In a poem entitled "Upon the First Knesset," Nathan Alterman beautifully defined the intriguing union between the ancient Hebrew term "Knesset" (which connotes the Great Knesset of the Second Temple era) and the name given to the first Israeli parliament, the "Founding Convention," a term borrowed from the formation of the French Republic. This marriage between Knesset Israel and the concepts of liberty and democracy is not a simple one. It's no wonder that voices in both camps have called to dissolve the match. But there can be no dissolution short of a total rift. And even if there could, it would be inadvisable. We should, rather, try to renew and heal this complicated couplehood.

The rift between Halachic Judaism and those

who do not live according to Jewish law is partial and not absolute. Any attempt to strengthen what the religious call "Jewish character" by means of coercion or legislation, or to hasten the coming of the Messiah by deploying tanks, or to force us all into religious observance by means of kashrut supervisors, will serve only to deepen the rift. Israel cannot be forcibly "Judaized." Let us assume that by some miracle the religious parties managed to enforce Halachic law on the entire state. Let us assume that no one moved a finger or a toe in all of Greater Israel in any way that contravened *Shulchan Aruch* or was not authorized by seven rabbis. Would such a state really be a more Jewish one? More infused with Jewish heritage? Or would it be more oppressed and desperate? More poisoned with frustration and anger?

Let us assume, conversely, a set of circumstances in which the Messianic zealots managed to destroy the Temple Mount mosques, rebuild the Holy Temple in their stead, and enlist a Knesset majority to outlaw the rest of the world. Let us say that they managed to annex all the occupied territories, eliminate all the Arabs once and for all, and cut Europe and America down to size. Would this make things better for the Jewish people? Or would it perhaps bring total devastation upon us, much as our zealots have done before, more than once?

THE CURRENT ISRAELI RIGHT WING AND Halachic Judaism share a common denominator: both are accustomed to living in real or emotional conflict with the outside world. Both maintain that there is an eternal clash between Israel and the nations of the world: "A sheep among seventy wolves,"* "Israel is a scattered sheep,"† "Like lambs to the slaughter."‡ The Israeli left wing's attempts to resolve the "eternal conflict" between us and the Arabs, between us and the whole world, are perceived by Halachic Judaism and much of the right wing as a dangerous threat to the uniqueness of the Jewish people: if there is no enemy,

---

* Midrash Tanchuma, Toldot, chap. 5.
† Jeremiah 50:17.
‡ Isaiah 53:7.

no persecution, no siege, and no "sanctification of God's name," then the outside world will seduce us and we will lose our identity and assimilate. In fact this outlook perceives the persecutions, the siege, the "enemy at the gate," as a "friendly" situation that we have been familiar with throughout our history, and which serves to reinforce our identity.[*]

In a normative condition there would be no token enemy facing us day and night (such a condition was envisioned in the Zionisms of Theodor Herzl and Chaim Weizmann, Ben-Gurion and Ze'ev Jabotinsky, to which the Israeli left wing still aspires). However, for many on the right and among Halachic Judaism and settler Judaism, such a condition threatens our very identity. "In every generation, they rise against us to annihilate us. But the Holy One, blessed be He, saves us from their hand" — this verse is recited every year on Passover. But if the day comes when they are no longer rising up to destroy us, what will our

---

[*] This topic has been discussed by Professor Dan Miron.

identity be? Whom will the Holy One save? There are quite a few Jews who believe that although the occupied territories are important to us, what is more important is that the eternal conflict continue to be eternal.

The divide between religious and secular Jews has existed for some 150 years. But it does not have to be a rift that beareth gall and wormwood; it can just be a rift that beareth. For this to happen, it is essential that there be mutual, genuine listening, not the affected listening of those who would convert secular Jews to religion, or the arrogant listening of those who have all the answers.

The tolerant and dynamic elements that Mizrahi Judaism brought to Israel, the Hebrew culture that sprung up with the first generations of Zionism, the new Hebrew literature, Yiddish culture and Ladino culture, the non-Orthodox streams of Judaism — all these present a challenging agenda for contemporary Israeli culture. Instead of squabbling over the superficial question of "who among us is more like my grandfather's grandfather?," we would do best to debate more substantive issues: What should we do with the heritage of all the generations? What stands in the center? What on the margins? What should we add? How should we add it? And how do we reject the outmoded elements?

WHEN THE SHOFAR WAS BLOWN AT THE Western Wall at the end of the Six-Day War, a genie was let out of the bottle. Since that day, Israelis both religious and secular, on the right and on the left, have obsessively debated the question of where the state's borders should run and which flag should fly over the holy sites. But for those who perceive them as such, those sites are inherently holy, flag or no flag. It is not the flag that bestows them with sanctity. The question of borders is indeed a weighty one, but only an obsessive considers it the most important question of all. What occurs *inside* the borders is exponentially more important than what their outline should be.

The State of Israel can be a monster or a cartoon caricature with expansive borders, just as it can be a fair, moral, creative society that fulfills its heritage and lives in peace with itself, inside narrow borders. It is madness to allow the question of borders to enslave and distort all other issues. This issue has never, in all of Jewish history, been the only one, or even the first one, on the agenda. We must finally awaken from the hypnosis of the map. It is time to talk about the fundamentals: What is going to happen here? Who will we be? Can we make another two or three hopes come true in this country, in addition to the ones that the State of Israel has realized and is still fulfilling?

HUNDREDS OF THOUSANDS OF JEWS IN ISrael and elsewhere, mostly young people, are increasingly seeing Judaism as a threatening branch of extremism, a sort of nationalist, belligerent, oppressive fist. Or an extortion operation. Or a steamroller that threatens their private lives and personal freedoms. These emotions endanger the most basic affinity of masses of Jews in Israel and abroad with Jewish heritage. This aversion makes people say, "Fine, those black hats can take their Judaism and leave me alone."

The Bible, the Mishnah and the Gemara, the liturgies, the prayer books, the poetry of Sepharad and Hasidism — this whole wonderful library now strikes many Jews as just another part of the steamroller coming to flatten them. More and more young people feel like throwing it all away and setting off in search of a completely different spirituality, perhaps in an ashram in India.

To me this is a grave tragedy. It is cause for a profound reckoning among the Haredim, the Messianic nationalists, and us secular Jews as well. It saddens me to see how self-deprecating many secular people become in the face of a supposed authenticity found among devout Jews. It's as though we're all supposed to acknowledge that the long-bearded rabbis in their black hats and black coats are the most Jewish Jews. The big league. In second place, ostensibly, stand minor-league Jews like the settlers, who may not have black hats and black coats, but at least they stick it to the Arabs and show them who's in charge. On the next rung down the ladder are the "traditional" Jews, those who sometimes maintain a little Yiddishkeit, fast on Yom Kippur (at least until noon), drive their cars on the Sabbath but do not eat pork. Further down, almost at the bottom, are the "commoners"—simple, well-meaning Jews who have lost their way, "captured infants," low-hanging fruit for the converters.

At the very bottom are the worst of the worst, the most un-Jewish, the most Israel-hating, the most goyish: the lefties, who insist on pursuing peace and protecting human rights, who won't let anyone quietly commit a minor injustice or a little nationalist usurpation, and who won't stop droning on about "Justice, justice shalt thou pursue,"[*] "Ye shall have one manner of law, as well for the stranger as for the home-born,"[†] "Thou shalt not kill," and "Seek peace and pursue it."[‡] If you so much as deport a few terrorists' relatives or deliver a collective punishment to a whole Arab village, those lefties start badgering us with foreign notions like "Every man shall be put to death for his

---

[*] Deuteronomy 16:20.

[†] Leviticus 24:22.

[‡] Psalms 34:16.

own sin."* Where on earth did these goyim come up with all those bleeding-heart concepts?

PERHAPS IT IS TIME FOR US ALL TO PAUSE and reopen the question of "Who is a Jew?" Not in the sense of whom the population registry considers Jewish, but rather, who among us is closer to that little potsherd from Khirbet Qeiyafa, and how many of us have long forgotten it?

Isaiah 55:8–9 says, "For My thoughts are not your thoughts, neither are your ways My ways, saith the Lord. For as the heavens are higher than the earth, so are My ways higher than your ways, and My thoughts than your thoughts." Those who would ignore these verses, those who think they are fulfilling God's precise will, are nothing but arrogant. We have no pope. And that is a good thing. We should all be cautious of anyone who purports to have deciphered a divine plan with a divine schedule and tries to fulfill it by political or military means. Many of us aspire to nurture and

---

* Deuteronomy 24:16.

enrich the Jewish heritage. But the essence of our heritage must not be equated with *Shulchan Aruch*, which has considerable value to Judaism but is not absolute. The Jewish people existed before *Shulchan Aruch*, and there is life outside it and after it. Moreover, we have other treasures that stand far above the collections of laws, such as that little potsherd. Or the prophets and their moral teachings.

Democracy and pluralism are simply popular expressions of the sanctity of life and the equality of human worth. They are manifestations of the Talmudic verse "Whoever saves one life, it is considered as if he saved an entire world." These notions are not foreign, not "imported." The sanctity of human life derives directly from the innermost core of the Jewish spirit at its finest. I believe that it is the very same place that gives rise to "Cause no pain" and to "That which is hateful to you, do not do to your fellow."

Secular Jews are also the heirs of Jewish culture. Not the only heirs, but perfectly legal ones. And heirs should not be enslaved to their inheritance. On the contrary: legal heirs are entitled to discard parts of their inheritance and to emphasize other parts, as they see fit. When I inherit my parents' house, I have the right to decide which furniture will be stored in the basement or the attic and which will remain in the living room. Obviously, the next generation is fully entitled to reverse the order of things: they may bring up from the basement or down from the attic whatever I removed, and refurnish the living room.

Rabbi Yonatan ben Yosef says of the Sabbath: "It is committed to your hands, not you to its hands."[*] In precisely this way we might describe the dispute between us and the Haredim, and between us and the Messianic settlers: we secular Jews believe that our past is committed to our hands. We are not committed to our past's hands. And that is more or less the heart of the disagreement: Does our past belong to us, or we to it? Roughly a century ago, Micha Josef Berdyczewski wrote: "The supreme law for Jews on Ju-

---

[*] Talmud, Tractate Yoma 85.

daism: the living man takes precedent over his patrimony."

What blossoms today may fertilize what will blossom tomorrow, and what blossoms tomorrow may resemble, conversely, that which blossomed yesterday. There are seasons in the life of a culture. For thousands of years, Jewish culture has absorbed pollinating seeds from other cultures, and has scattered its own stamen in different worlds. All this, to me, is enfolded in one half of one biblical verse: "Renew our days as of old."*

It is not possible to renew without days of old, and days of old cannot exist without renewal.

---

* Lamentations 5:21.

III

Dreams Israel Should Let Go of Soon

I BEGIN WITH THE MOST IMPORTANT MAT-
ter, a question of life or death for the State of Israel. If there are not two states here, very soon,
there will be one. If there is one state, it will be
an Arab one that stretches from the Mediterranean to the Jordan River. Jews and Arabs can and
should live together, but I would find it absolutely
unacceptable to be part of a Jewish minority under Arab rule, because almost all the Arab regimes
in the Middle East oppress and humiliate their
minorities. And more importantly, because I insist
on the right of Israeli Jews, like any other people,
to be a majority, if only on a tiny strip of land.

I said we would have an Arab state from the Mediterranean to the Jordan; I did not say it would be a binational state. Apart from Switzerland, all binational and multinational states are either barely squeaking by (Belgium, the United Kingdom, Spain) or have already deteriorated into violent conflict (Lebanon, Cyprus, the former Yugoslavia and USSR).

If we don't have two states, it is likely that in order to thwart the establishment of an Arab state from the Mediterranean to the Jordan, a temporary dictatorship will be instituted by fanatic Jews, a racist regime that uses an iron fist to oppress both Arab residents and its Jewish opponents. This sort of dictatorship will not last long. Virtually no minority dictatorship oppressing a majority has endured in modern history. And at the end of that road, too, what awaits us is an Arab state from the Mediterranean to the Jordan, possibly preceded by an international boycott, a bloodbath, or both.

Various self-appointed authorities insist that there is no solution to the conflict, and so they advocate "conflict management," which would mean a repeat of the cycle of wars along the northern and southern borders of Israel over the past few decades. Conflict management implies a series that begins with the first Lebanon war and the second Lebanon war and goes on to the third, fourth, and fifth. It implies the military operations in Gaza known as "Cast Lead," "Pillar of Defense," and "Protective Edge," to which we might have to add "Drawn Bow," "Iron Boots," and "Bloody Fists." And perhaps another intifada or two in Jerusalem and the West Bank, until we witness the collapse of the Palestinian Authority and the rise of Hamas — if not an even more radical entity. That is the meaning of "conflict management."

For more than a century ("one hundred years of solitude") we have not had a more favorable hour to end the conflict. Not because the Arabs have suddenly become Zionists, not because they are willing to recognize our historic right to the land, but because Egypt, Jordan, Saudi Arabia, the Emirates, the Maghreb countries, and even Assad's Syria have a far more immediate, more destructive, and more dangerous enemy than Israel.

More than a decade ago, in 2002, the Saudi peace plan, which was in fact the Arab League's plan, landed on our desk. I do not recommend rushing to sign on the dotted line at the end of that document, but it is certainly worthy of negotiating. We should have done so years ago, and perhaps then we would be in a completely different situation. If we had received a similar proposal in the days of Ben-Gurion and Levi Eshkol, in the "no" days of the Arab League summit in Khartoum, in 1967, all of us would have been dancing in the streets.

Here I will make a controversial statement: since the Six-Day War in 1967, Israel has not won a single war. Including the Yom Kippur War. War is not a basketball game in which whoever gets more points wins the trophy and a handshake. In war, even if we burned more tanks than the enemy did, or shot down more planes, or occupied more territory, or killed more of them than they did of us, it does not mean we won. The winner in war is the side that achieves its objectives, and the loser is the one that does not. In the Yom Kippur War, Anwar Sadat's objective was to shatter the status quo that had evolved since the Six-Day War, and in that he succeeded. We were beaten because we did not achieve our objective, and we did not achieve it because we had no objective, nor could we have had any objective that could be obtained by employing military might. (This has been pointed out by my friend the late General Israel Tal.)

Am I saying that military might is unnecessary? Absolutely not. At any given time in the past several decades, our military strength has stood as a constant barrier between us and ruination. But we must bear in mind that when it comes to Israel and its neighbors, our military power can only prevent. It can prevent a catastrophe, prevent annihilation, prevent large-scale casualties among our population. But we cannot win a war, because we have no definitive national goals that can be achieved through military might. And so, as I said,

I view the notion of conflict management as a recipe for one trouble after another — not to mention defeat after defeat.

A great number of Israelis believe — or are brainwashed into believing — that if we simply carry a big stick and deliver just one more severe blow to the Arabs, they will finally get scared and leave us alone and everything will be fine. But the Arabs have not left us alone for almost a hundred years, despite our big stick. The quarrel between Israel and Palestine is a wound that has been bleeding for decades. A festering wound. There is no point in waving a big stick over and over again and beating the bloody wound so that it gets scared and finally stops being a wound and bleeds no longer. An open wound must be healed, and there is only one way to gradually heal this particular wound.

Meanwhile, the oppressive Israeli regime in the occupied territories is bringing down the Palestinian Authority. And when it comes tumbling down, we will find ourselves facing Hamas in the West Bank as well — if not worse. Millions of Palestinians in the territories live a life of constant humiliation, enslaved and denied their rights. Their human and national dignity are trampled, their property is forfeited, and their very lives under Israeli rule are cheap. About a third of the West Bank's lands have been robbed by Israel, and the robbery continues.

The right wing and the settlers claim we have the right to the entire Land of Israel and to the Temple Mount. But what do they actually mean when they say "right"? A right is not "something I really want and feel very strongly that I deserve." A right is what *others* recognize as such. If others do not acknowledge my right, or if only some of them acknowledge it, or only partially acknowledge it, then what I have is not a *right* but a *demand*. And that is exactly the difference between Ramleh and Ramallah, between Haifa and Nablus, between Be'er Sheva and Hebron. The whole world, including most of the Arab and Muslim powers (apart from Hamas, Hezbollah, and Iran) now recognizes that Haifa and Be'er Sheva are part of Israel. No one, however, except the settlers and their supporters in the American radical Christian right, recognizes that Nablus and Ramallah belong to us. That is the difference between a right and a demand.

The settlers and their supporters say, "We have the right to all of Eretz Yisrael." But in fact they mean something different: to them it is not just a right but a duty, a religious obligation, to possess every corner of the biblical land. When I stand before a pedestrian crossing, of course I have the right to cross the street. But if I see a truck hurtling toward me at sixty miles an hour, I am entitled not to exercise my right to cross the street. Our zealots are convinced that our religious duty is to ignore the dangers and cross the street, and that God will be on our side.

Take, for example, the Temple Mount. Of course Jews have the right to pray there. But we also have the prerogative to choose not to exercise that right — at least not in this generation. There are those among us who are no longer impressed by the conflict with two hundred million Arabs; they're bored with it, they're sick of it, they want action.

They want to lead us into a war with all of Islam. With Indonesia and Malaysia, with Iran and Turkey, with nuclear Pakistan. Is it really worth dying for prayers on the Temple Mount? There is no such commandment in all the Jewish sources. The aspiration to pray on the Temple Mount is not a life-or-death matter, not even according to Jewish law. If you want to ignite a world war against all of Islam for the sake of the Temple Mount, go ahead and do it without me, please, and without my children. Without my grandchildren.

But these elements are not satisfied with waging war against all of Islam. There are those who are trying to lead us into war against the whole world. Some forty years ago, the day after the 1977 elections that brought Likud to power for the first time, one newspaper editor was so overjoyed at the reversal that he began his front-page editorial with the following words: "Likud's victory in the Israeli election cuts America down to size." Today we are still witnessing an Israeli attempt to cut the whole world down to size. To destroy the alliance between most of the American people and Israel, or to replace it with a narrow alliance between our extreme right wing and theirs. During the months-long struggle over the future of the Iranian nuclear program, there were those who said something like this: "The leader of the free world is fighting alone against Iranian nuclear weapons. Why does President Obama keep getting in his way?"

The election of Donald Trump as president of the United States has exhilarated the settlers and their supporters, as though they had managed to erect a new settlement in the heart of the White House. But it is worth remembering that most Americans voted against Trump, and also that most Trump supporters did not vote for him because they love the idea of a Greater Jewish Israel. In fact, President Trump and his administration do not support the annexation of the occupied territories by Israel and the creation of an expanded state in which the Jews live upstairs and the Palestinians live downstairs in the servants' quarters. Apart from the settlers and the American Christian zealots who back them, the whole world is united in opposing Israel's taking over the West Bank lands and controlling their Palestinian residents.

WE MUST NOT FORGET THAT AT LEAST twice in our history we have found ourselves in a head-on collision with a large superpower. Both times — when we rose against Babylonia and against Rome — things ended badly for us.

I see days not far ahead when mechanics in Amsterdam, Dublin, or Madrid refuse to service El Al airplanes. Consumers boycott Israeli products. Investors and tourists stay away from Israel. The Israeli economy collapses. We are already at least halfway there.

David Ben-Gurion taught us that the State of Israel cannot survive without the support of at least one superpower. Which one? That changes: first it was Great Britain, for a time it was Stalinist Russia, for one brief window it was England and France, and in the past few decades — the United States. But the alliance with America is certainly not a law of nature or a permanent state of affairs.

. . .

ONE OF THE MOST IMPORTANT DISTINC-
tions in the lives of human beings, and of nations,
is between the permanent and the transient. For
decades, right-wing fear-mongers have been pre-
dicting that if we give back the territories, "communist tanks from the Warsaw Pact will turn up
outside Kfar Saba." No one can guarantee that if
we withdraw from the territories everything will
be wonderful, but we can say for sure that there
will be no communist tanks near Kfar Saba. We
must not confuse a transient factor with a permanent factor.

The same serial alarmists who scared us with
the Red Army are now threatening that if we withdraw from the territories, rockets will fall on Tel
Aviv, Ben Gurion Airport, and Kfar Saba. Whether
or not that is true, I can say with the full authority
of a staff sergeant in the IDF that those sites can
already be hit, not only by rockets launched from
Kalkiliya in the West Bank, but also by ones fired
from Iraq, Pakistan, and perhaps Indonesia. Again
we have a miserable failure to distinguish the mu-

table from the unchanging. If not today, then certainly tomorrow or the next day, it will be possible to launch rockets from any point in the world to any other point, achieving a precise and lethal hit. Are we going to dispatch the IDF to conquer the whole planet?

The fact that the United States is our ally is an impermanent fact. It might change. But the fact that the Palestinians are our neighbors and that we live in the heart of the Arab and Muslim world — these are both permanent conditions. The prevalent Israeli perception, whereby the nuclear threat originates solely from Iran and that it can be defeated by destroying the Iranian nuclear facilities, is one that does not distinguish the passing from the permanent, because even if we or someone else were to bomb the Iranians' nuclear plants, we could not also bomb the knowledge they possess. Moreover, nuclear Pakistan could quickly become an Islamist state more radical than Iran. There is nothing to prevent Israel's wealthy enemies from buying ready-made nuclear weapons and aiming them at us. In a number of years, anyone who wants weapons of mass destruction will be able to obtain them. On this matter, too, it is essential to learn how to distinguish the temporary from the permanent. The permanent factor must be Israel's deterrence force, whereas our enemies' capacities, nuclear and otherwise, are a changing matter that does not, ultimately, depend on us.

I have said that unlike some of my friends on the dovish left, I cannot guarantee that if we sign a peace accord and withdraw from the territories, everything will be rosy. But I am sure that if we stay there, things will be worse. If we continue to occupy the territories, there will eventually be an Arab state from the Mediterranean to the Jordan. On this point some members of the dovish left, myself included, deserve criticism. There are millions of Israelis who might be willing to give up land for peace, but they do not believe the Arabs. They do not want to be "suckers." They're afraid. No one should ever diminish or mock this fear, although we can try to untangle it. To assuage it. And it might not be bad for the Israeli dovish left to partake in that fear, just a little.

Because there are genuine reasons for fear. But either way, a person who is fearful, whether justifiably or not, should never be mocked or belittled. The concept of peace for land should be debated without mockery or dismissal, weighing one danger against another.

Another mistake made by some of my friends on the dovish left is that sometimes they act as though peace were sitting high up on a shelf in a toy store: one has only to reach up and touch it. Yitzhak Rabin almost touched it at the Oslo Accords, but he was too tightfisted to pay the full price, and so he came home without buying us the toy. Ehud Barak almost touched it at Camp David, but he was also too stingy, so he came home empty-handed as well. The same happened with Ehud Olmert — another daddy who was penny-pinching, who didn't love us enough, because otherwise he would have paid the full price and brought us peace long ago.

I do not accept that view. I believe that peace must have more than one partner. There is an Arab saying that goes, "You can't applaud with one hand." But today we do have a partner for negotiations. The brainwashers told us for years that Yasser Arafat was too strong and too malicious, and now they tell us that Mahmoud Abbas (also known as Abu Mazen) is too weak. They tell us that as long as the Palestinians keep killing us, we can't make peace with them. But when they stop killing us, there's no reason to make peace with them.

My Zionist point of departure, for decades now, has been a simple one: We are not alone in this land. We are not alone in Jerusalem. I say the same thing to my Palestinian friends: You are not alone in this land. There is no escape from dividing up this little house into two even smaller apartments. Into a duplex. If someone from either side of the Israeli-Palestinian divide comes along and says, "This is my land," he is right. But if someone says, "This land, from the Mediterranean Sea to the Jordan River, is *all* mine and *only* mine," then he is out for blood.

There must be compromise between Israel and Palestine. There must be two states. We must divide this land and turn it into a duplex.

On both sides there are many people who loathe the very idea of compromise, viewing any concession as weakness, as pitiful surrender. Whereas I think that in the lives of families, neighbors, and nations, choosing to compromise is in fact choosing life. The opposite of compromise is not pride or integrity or idealism. The opposite of compromise is fanaticism and death.

The Palestinians are essentially waging two different wars with us. On the one hand, many of them fight to end the occupation and for their just right to national independence, to be "a free people in its land," in the words of our own national anthem. Every decent person must support such a struggle, albeit not all the means the Palestinians use. On the other hand, many Palestinians are waging a war of fanatical Islam, a war for their fervent aspiration to demolish Israel as the state of the Jewish people and the state of all its citizens. (According to fanatical Islam, the Jews are too despicable to be considered a nation, and at most are allowed to pray in synagogues and exist as humble individuals under the "protection" of Islamic rule, much as the Jews of Yemen, Iraq, Iran, and Syria existed for many generations.) That is a criminal war that any decent person must resist.

The source of confusion and oversimplification — among us Israelis and around the world — is that many Palestinians are waging both these wars at the same time. Even decent people who strive for peace and justice, in Israel and elsewhere, fall into the trap. Either they zealously defend the continued occupation of the territories, claiming that Israel is a victim of jihad and that the occupation is an act of self-defense, or they denigrate Israel and insist that the occupation is the sole source of evil, and that the Palestinians are therefore entitled to shed Israeli blood.

Two wars are being waged: one is exceedingly just and the other entirely unjust. It's Dr. Jekyll and Mr. Hyde. The State of Israel is also Jekyll and Hyde, fighting two wars simultaneously: a just war for the Jewish people's right to be a free nation in its own land, and a war of oppression, injustice, and theft meant to add another two or three rooms to our apartment at the expense of our Palestinian neighbors, to rob their lands and deny their right to liberty.

The idea of a binational state, which has gained support on the extreme left as well as from some figures on the delusional right, I regard as a sad joke. We cannot expect Israelis and Palestinians, after a hundred years of blood, tears, and catastrophes, to jump into a double bed together and begin their honeymoon. If someone had proposed in 1945 that Poland and Germany unite into one binational state, he would have undoubtedly been institutionalized.

A few days after the great victory in the Six-Day War, I was already writing about the "total moral devastation that prolonged occupation engenders among the occupier." And even then, I feared that the occupation would also corrupt the occupied people. Many fine thinkers, among them Yeshayahu Leibowitz, shared my view.

The Palestinians and Israelis cannot turn into a happy family overnight. We need two states. Sometime later there might come cooperation, a joint market, a federation. But first the country must be a duplex, because we Jewish Israelis are not going anywhere. We have nowhere to go. The Palestinians are not going anywhere either. They, too, have nowhere to go. The quarrel between us, fundamentally, is not a Hollywood Western pitting good against bad, but a tragedy of justice against justice. That is what I wrote almost fifty years ago, and I still believe it today. Justice against justice — and often, to my sorrow, injustice against injustice.

When a patient is rushed into an operating theater with multiple injuries, the surgeon asks himself: What first? What is most urgent? What might kill the patient? In the case of Israel, the primary danger is its continued conflict with the Arabs, which is destined to become a conflict with most of the world's countries and endanger our very existence.

Perhaps this is the place to expose Israel's most closely held security secret. The secret is that we are in fact weaker, and always have been, than the sum of our enemies. Our enemies have been brainwashed for decades with wild rhetoric about destroying Israel and throwing the Jews into the sea. But what has prevented them from dispatching a million troops to fight us — or two or three million? They have never sent more than a few tens of thousands, because despite the murderous rhetoric, Israel's existence has never been a life-or-death matter for Syria, nor for Libya, Egypt, or Iran. Perhaps it has for the Palestinians, but fortunately for us they are too few and too weak to overcome us. A confederation of our enemies could have long ago destroyed us if, God forbid, they had genuine motivation and not just rhetoric and propaganda. Our next adventure on the Temple Mount might create that motivation.

I am not sure that we can end the fight with the Arabs overnight. But we can try. I believe that it was possible a long time ago to reduce the Israeli-Palestinian conflict to an Israeli-Gazan one.

IT IS HARD TO BE A PROPHET IN THE LAND of prophets. There's too much competition. But my life experience has taught me that in the Middle East, the words "forever," "never," and "not for any price" may mean anything between six months and thirty years. If I had been told, when I was called up for reserve duty in the Sinai Desert during the Six-Day War, and again on the Golan Heights during the Yom Kippur War, that one day I would visit Egypt and Jordan with my Israeli passport, I, the dove, the peace-monger, would have said: Don't get carried away. Maybe my children or my grandchildren will, but not me.

THE WORD THAT MOST IRRITATES AND outrages me these days is "irreversible." The extreme right wing on the one hand, and the post-Zionist and anti-Zionist groups in Israel and the world on the other, seem to have entered a secret pact, a conspiracy, to brainwash us with the assertion that the Israeli occupation of the territories is irreversible, the settlements are irreversible, the chance for a two-state solution is irreversibly lost. The fanatical right uses the word "irreversible" to tell us that Israel's annexation of the territories is final and absolute, and if we do not want to live under an Arab-majority rule between the Mediterranean and the Jordan River, we must forget about democracy. We must accept that the Jews will rule Greater Israel and the Arabs will be content to perform menial labor. Our fanatics tell us, not without a measure of gloating: You Jewish Israelis who value democracy so much, from now on you'll have to forget about it. If you don't want to live in an Arab state under Arab rule, you'll have to give up your beloved democracy, the rule of law, the Supreme Court, and get used to living under the Jewish rule of the hilltop rabbis. Don't like it? Get out and go wherever.

The post-Zionists and the anti-Zionists, from Tel Aviv all the way to college campuses in the West, also repeat the refrain that the occupation is irreversible and that, just like South Africa, we must choose between living as a nation of masters in a Jewish apartheid state, and giving up the Zionist dream and accepting our fate as a minority under Arab rule. Don't like it? You're free to leave.

This dual brainwashing—from the left and the right—about the irreversibility of the situation is meant to break the spirit of the Zionist left, which opposes the occupation and refuses to rule over another nation, yet still believes that the Jewish people have a natural, historical, legal right to a sovereign existence as a majority, if only in a very small democratic state. The Zionist left is despised by the hilltop settlers on the one hand, and by the post-Zionist and anti-Zionist front on the other. They have both been denouncing the left for years and are eager to trounce it. That is why it sometimes seems that these two extremes have conspired to make us despair and force us to choose between giving up on Zionism and giving up on democracy. They hope the choice will be so intolerable that we will simply get up and leave.

BUT WHAT ON EARTH DOES THE WORD "irreversible" mean? What is irreversible about Israel's occupation and its oppressive control of the Palestinian territories?

Those who have witnessed with their own eyes, as I have, the founding of the State of Israel, only three years after the murder of European Jewry by the Nazi Germans, will be in no hurry to ac-

cept the term "irreversible." Those who have seen
Charles de Gaulle, hero of the French right wing,
grant independence to Algeria, which was an-
nexed by France and populated by hundreds of
thousands of fanatic French settlers, will be in
no rush to take the threats of the "irreversible"
prophets at face value. Those who have seen the
peace-seeking Levi Eshkol, after the Six-Day War,
rule over the most expansive Jewish state since
King David's era; who watched Menachem Begin, leader of the Israeli right, dismantle Eshkol's
empire exactly ten years later in order to sign a
peace accord with Egypt; who beheld President
Sadat, "King of the Arabs" and Israel's chief enemy, stand at the Knesset podium to offer peace
and recognition in return for territories, and Menachem Begin accept his outstretched hand; who
witnessed John F. Kennedy, hero of the American
left, embroil the United States in the mire of Viet-
nam, and the right-wing Richard Nixon extricate
it from that war; who saw Yitzhak Rabin and Shi-
mon Peres, both hawks who supported the settlements, shake Yasser Arafat's hand and try to ac-

complish a two-state compromise with him; who saw Komsomol cadet Mikhail Gorbachev dismantle the Soviet empire; who watched Ariel Sharon's bulldozers raze Ariel Sharon's settlements in Gaza — those who have seen all that and more will not easily buy into the despair potion known as "the irreversible situation." I maintain, moreover, that the fanaticism of the radical settlers may itself be reversible. The dogmatism of the anti-Zionist left might be, too. Perhaps the only irreversible thing is death, and even that will have to be examined closely one day, in a very personal way.

In conclusion, we are permitted to briefly turn our gaze away from the existential fear, to the fact that for decades now there has been a cultural golden age in Israel, in literature, cinema, music, theater, fine arts, philosophy, science, and technology. People usually wax nostalgic for a golden age only after it has passed. But Israel has been at the height of such a creative era for decades. The city of Tel Aviv, for example, known as the first Hebrew city, is to me a collective Israeli creation no less important, and perhaps more so, than the rabbinical literature written in the Diaspora or the codification of Jewish law contained in *Shulchan Aruch*. Tel Aviv is only one of Israel's creations among many other cities and towns, villages, moshavim, and kibbutzim.

There are those who view this new Hebrew culture as too leftist. There have been — and still are — regimes that incite against culture, mostly because the majority of cultural figures almost always have an anti-authoritarian bent. In different eras and disparate countries, however, many artists have in fact opposed the regime from the right rather than from the left. That was the case in Israel about fifty years ago, under the left-wing government, when almost all the notable authors joined the Greater Israel movement, which stood on the farthest right end of the political spectrum.

Now comes a little confession: I love Israel even when I cannot stand it. If I have to fall over in the street one day, I would like it to happen on a street in Israel. Not in London, not in Paris, not in Berlin or New York. Here people will come over immediately and pick me up. (Granted, once I'm back on my feet there will probably be quite a few who will be happy to see me fall down again.)

I am extremely fearful for the future. I fear the government's policies, and I am ashamed of them. I am afraid of the fanaticism and the violence, which are becoming increasingly prevalent in Israel, and I am also ashamed of them. But I like being Israeli. I like being a citizen of a country where there are eight and a half million prime ministers, eight and a half million prophets, eight and a half million messiahs. Each of us has our own personal formula for redemption, or at least for a solution. Everyone shouts, and few listen. It's never boring here. It is vexing, galling, disappointing, sometimes frustrating and infuriating, but almost always fascinating and exciting. What I have seen here in my lifetime is far less, yet also far more, than what my parents and their parents ever dreamed of.

Acknowledgments

Many people read drafts of this book, or of parts of it, at various stages of the writing and contributed their wisdom, expertise, and sensitivity. Warm thanks to everyone who improved the text of this little book:

Nily Oz, Fania Oz-Salzberger, Eli Salzberger, and Daniel Oz.

Deborah Owen, Haim Oron, Emuna Elon, Charles Buchan, Haim Be'er, Ilan Bar David, Nahum Barnea, Uzi Baram, Tzvia Glezerman, David Grossman, Dalit Gertz, Nurit Gertz, Eran Dolev, Avner Holtzman, Andrew Wylie, Avinoam Werbner, Miri and David Varon, Dorit Zilberman, Shira Hadad, Shai Huldai, David Chen,

Gadi Taub, Dalia Yairi, A. B. Yehoshua, Menahem Yaari, Nili Cohen, Rabbi Binyamin Lau, Dan Laor, Niva Lanir, Gafnit Lasri Kokia, Avishai Margalit, Einat Niv, Noa and Yoram Amit, Muki Tsur, Nissim Kalderon, Aviad Kleinberg, Aliza Raz-Melzer, Tzali Reshef, Yigal Schwartz, Youval Shimoni, Avraham Shapira, and Anita Shapira.

Bibliography

I. Dear Zealots

Amichai, Yehuda. "The Place Where We Are Right." In *The Selected Poetry of Yehuda Amichai,* edited and translated by Chana Bloch and Stephen Mitchell. University of California Press, 1996.

Hoffer, Eric. *The True Believer: Thoughts on the Nature of Mass Movements.* Harper, 1951.

Oz, Amos. *Panther in the Basement,* translated by Nicholas de Lange. Harcourt Brace, 1997.

Oz. Galia. *Rebelling Against the Kingdom* (documentary film). Channel One, 2007.

II. Many Lights, Not One Light

Agnon, S. Y. (Shmuel Yosef) *A Guest for the Night.* Schocken, 1939.

Alterman, Nathan. "Upon the First Knesset." In *Ha'Tur ha'shvi'i,* book 2, p. 44. Davar, 1953.

Amichai, Yehuda. "God Full of Mercy." In *The Poetry of Ye-*

huda Amichai, translated and edited by Robert Alter. Farrar, Straus & Giroux, 2015.

Berdyczewski, M. Y. (Micha Yosef) "Conflict and Construction." In *The Complete Essays of Micah Yosef Ben Gurion (Berdyczewski),* pp. 29–30. Amo Oved, 1952.

Bialik, Chaim Nachman. "On the Slaughter," translated by Peter Cole. *Paris Review,* July 31, 2014.

Greenberg, Uri Zvi. "At the End of Roads Stands Rabbi Levi Yitzhak of Berdichev and Demands a Great Answer." In *Rehovot Ha'Nahar* ("Streets of the River"). Schocken, 1951.

Yizhar, S. (Yizhar Smilansky) *Khirbet Khizeh,* translated by Nicholas de Lange and Yaacob Dweck. Ibis Editions, 2008.

Yizhar, S. "The Courage to Be Secular," *Shdamot* 79, 1981.

Zemach, Shlomo. *Essays and Criticism,* p. 35. Dvir Authors Society, 1954. The precise wording of Zemach's analysis of Brenner's protagonists is: "His protagonists protest before God — as a sign that they have God in their hearts."

III. Dreams Israel Should Let Go of Soon

Oz, Amos. "The Minister of Defense and Living Space." In *Under This Blazing Light,* translated by Nicholas de Lange. Canto, 1996.